The Making of Hominology

A Science Whose Time Has Come

BOOK TESTIMONIALS

Dr. Jane Goodall

Dmitri Bayanov has proposed a new scientific discipline—hominology—that will study the many reports of unclassified hairy upright hominoid-like creatures from various quarters of the globe. Beginning in the 1960s, Bayanov worked directly with Professor Boris Porshnev and other Russian scientists investigating reports of relict hominoids, such as the *almasty,* described as a possibly extant Neanderthal. Continuing that work, Bayanov has authored several books and published papers arguing convincingly that the accumulating evidence for these species warrants a move from the realm of myth and legend to serious scientific investigation. A lifetime of scholarly examination of this question, with evidence spanning from the dawn of written communications to the present, has culminated in this important book – *The Making of Hominology.*

Jane Goodall, PhD, DBE; Founder – the Jane Goodall Institute & UN Messenger of Peace

Dr. Nikolay Drozdov

Serious study of the "snowman" phenomenon began with Prof. Boris Porshnev's groundbreaking book *The Present State of the Question of Relict Hominoids* (1963). He noted in it the emerging science of still unclassified higher bipedal primates that later developed into a discipline termed hominology. As a Moscow University student I attended Prof. Porshnev's eye-opening lectures on this subject and was presented by him a copy of his famous book. The relevant research had been initiated by him at the Academy of Sciences and continued at the Darwin Museum by a group of enthusiasts, headed first by museum Chief Curator Pyotr Smolin and followed by Dmitri Bayanov. His present book *The Making of Hominology,* written in association with Christopher Murphy, is a timely and substantial contribution in this frontier of scientific investigation.

Nikolay Drozdov, PhD, Doctor of Sciences in Biology; Doctor in Geography – Chair of Biogeography, Lomonosov Moscow University, Russia

Dr. Paul LeBlond

For years Dmitri Bayanov has argued forcefully for a scientific approach to the interpretation of the evidence for wild hominids (Sasquatch, Yeti, Almasty…). This book is an eloquent summary of his struggle to promote a scientific "hominology." It also provides examples of the sober and detailed examinations which he advocates, applied to some of the available evidence. A serious and thoughtful book on a controversial subject.

Paul LeBlond, PhD, Professor emeritus, Dept. Physics and Oceanography, University of British Columbia

Dr. Henry Bauer

This book makes the explicit case that the study of yetis, Sasquatch, and the like qualifies as a science—hominology—both because of the nature of science and because the evidence is overwhelming that these creatures are real—and that they are closer relatives of humans than of apes. The author has been with this project essentially from the beginning, and his accounts of its history are authentic. A valuable resource for both fans and skeptics.

Henry Bauer, PhD, Professor emeritus of Chemistry and Science Studies, University of Sydney, Australia

THE FOUNDERS OF HOMINOLOGY

The founders of hominoid research in Russia: (left to right) Boris Porshnev, Alexander Mashkovtsev, Pyotr Smolin, Dmitri Bayanov, and Marie-Jeanne Koffmann. The photograph from which these images were obtained was taken in January 1968. Boris Porshnev arranged a photographer, invited his very close friends and colleagues and said, "This is for us to be remembered in the future." (© D. Bayanov)

The Making of Hominology
A Science Whose Time Has Come

by

Dmitri Bayanov
In association with Christopher L. Murphy

Edited by Christopher L. Murphy

ISBN-13: 978-0-88839-011-0 *[trade paperback]*
ISBN-13: 978-0-88839-285-5 *[trade hardback]*
ISBN-13: 978-0-88839-289-3 *[epub format]*
Copyright © 2019 Dmitri Bayanov

Library and Archives Canada Cataloguing in Publication

Bayanov, Dmitri, author.
    The Making of Hominology: *A Science Whose Time Has Come* / by
Dmitri Bayanov in association with Christopher L. Murphy; edited by
Christopher L. Murphy.

Issued in print and electronic formats.
ISBN 978-0-88839-011-0 (softcover)—ISBN 978-0-88839-012-7 (PDF)
    1. Sasquatch. 2. Primates. I. Murphy, Christopher L.
(Christopher Leo), 1941- editor II. Title.

QL89.2.S2B436 2017   001.944   C2017-905905-X
                                           C2017-905906-8

Editor: Christopher L. Murphy
Book design: Christopher L. Murphy
Front cover image: Wild man with shield by Martin Schongauer, 1490.

Printed in the USA
Published simultaneously in Canada and the United States by:

HANCOCK HOUSE PUBLISHERS LTD.
19313 Zero Avenue, Surrey, BC, Canada V3Z 9R9
(604) 538-1114   Fax (604) 538-2262
HANCOCK HOUSE PUBLISHERS
#104-4550 Birch Bay-Lynden Road, Blaine, WA, USA, 98230
(800) 938-1114   Fax (800) 983-2262
www.hancockhouse.com   sales@hancockhouse.com

CONTENTS

Notations:
—The Chapter header images show a sculpture created by Alexandra Bourtseva in 1974, based on the Patterson and Gimlin film subject. It was gifted to Dmitri Bayanov in that year.
—The Patterson and Gimlin film is also shown as the Patterson-Gimlin film and may be abbreviated as PG film or PGF as applicable.
—The words "Sasquatch" and "Bigfoot" are not spelt with a capital letter if quoted text or the original document has them without a capital.
—Live links for referenced material are in the ebook provided by Hancock House Publishers.
—The use of the symbol /.../ means words or sentences that have been omitted for brevity.

ACKNOWLEDGMENTS

A list of my fellow researchers throughout the world would be too extensive for this volume, so I must simply express my gratitude to all and show only those who have been highly significant in my research.

NORTH AMERICA

Roger Patterson
Bob Gimlin
Ivan Sanderson
John Green
René Dahinden
Dr. Grover Krantz
Peter Byrne
Dr. Henner Fahrenbach
George Haas
Dr. Jeff Meldrum
Bobbie Short
Chris Murphy
Will Duncan
Dr. John Bindernagel

RUSSIA

Dr. Boris Porshnev
Dr. Alexander Mashkovtsev
Pyotr Smolin
Dr. Marie-Jeanne Koffmann
Dr. Igor Burtsev
Alexandra Bourtseva
Dr. Dmitri Donskoy
Dr. Michael Trachtengerts
Vadim Makarov

ABOUT THE AUTHORS

Dmitri Bayanov was born in Moscow, Russia, in 1932. He went on to become one of the foremost Russian cryptozoologists and hominologists alive today. After studying under such individuals as Professor B.F. Porshnev and P.P. Smolin, chief curator of the Darwin Museum in Moscow, Dmitri took part in Marie-Jeanne Koffmann's expedition in search of the Russian snowman in the Caucasus and made reconnaissance trips in the same region on his own (1970s). He was a founding board member of the International Society of Cryptozoology and served on its Board of Directors until 1992. Dmitri is currently an active member of the Relict Hominoid Research Seminar at the Darwin Museum (since 1964) and became its chairman in 1975.

Dmitri is credited with coining the terms "hominology" and "hominologist" in the early 1970s to describe the specific study of unknown hominoids and those who study them. Dmitri's hominological career has been spent mainly on the study of relict populations of hominoids including the Russian snowman and the North American Sasquatch or Bigfoot. This is reflected in his several books published in Russia and Canada. He currently lives in Moscow, Russia.

Christopher L. Murphy was born near Mere, England, in 1941. He came with his parents to Vancouver, BC, Canada, in 1947. He entered the work force in 1957 and in 1958 went to work in the purchasing department for the B.C. Telephone Company (now Telus). He was promoted to management in 1969 and served in several management positions until retirement in 1994.

He became interested is sasquatch studies in 1993 upon meeting René Dahinden. He went on to write and co-write books on the subject and edited books for others. He collected sasquatch-related artifacts/artwork and was invited to provide an exhibit for the Museum of Vancouver, held in 2004/05. His exhibit has since traveled to seven other public museums in BC, Idaho, Oregon, Washington State, and Indiana. He has given talks on the subject at various conferences and has appeared in TV documentaries. He is a major contributor to the Sasquatch Canada website. Chris currently lives in Burnaby, BC, Canada.

FOREWORD BY DAVID HANCOCK

I was apprehensive to publish the first Hancock House book on sasquatch back in about 1975. I had just spent 10 years associating with scientists and going through graduate school and then nearly another 10 years starting a publishing company, focusing on down-to-earth history and natural history books. How could I take our "rationally based publishing company" and do books on a "fringe creature"—the mythical sasquatch? Well, it was those same roots that demanded it.

Charles Guiquez, Museum Biologist, Dr. Clifford Carl, Museum Director, Wilson Duff and Don Abbott, Museum Anthropologists, Dr. Adam Schawinski, Museum Botanist, York Edwards, Park Biologist, and a whole host of "drop-by-biologists" of Parks, Wildlife and Fisheries would share the tall round coffee table; the only place in the room where coffee cups, skulls, a bottled frog, ancient masks or dried plants ever came together. This hallowed spot was in the bowels of the old Provincial Museum (now the Royal Museum), where the discussions were a wondrous treasury of British Columbia wildlife and management theories.

Even the guru of wildlife Dr. Ian McTaggart-Cowan would drop by to feel the pulse of what was driving British Columbia wildlife studies. At the center of every table sat Frank Beebe, the most read, most consulted and most respected member of the Museum staff—the artist who never went to high school and whom all scientists depended upon to ground and give respectability and biological reference and representative illustrations to all their papers.

Frank became my mentor for the next 60 years. He taught me how to train a Cooper's hawk when I was 11 years old and led me on my first and many Museum and private expeditions. I remember an early paper he evolved at that coffee table in the Museum basement, the "Ecology of Sasquatch." Its evolution was inspiring. It partly evolved from the skepticism and arguments of those gathered biologists and yet often from the free-floated ridicule that were collectively offered. An often invited quest and stimulator of the topic at that round table was John Green, who emerged from the tunnels that permeated under the Museum and Parliament Buildings. He had been searching long-stowed boxes for old bones of sasquatches. I will not delineate who pooh-poohed his efforts or who supported them; I rather state that the consistent banter that always arose fostered my perennial interest in science and scientists—and sasquatches.

It was those conversations around that table that demanded, when I had completed my commercial pilots license, that I should go to university and become a biologist. It was those biologists, anthropologists and botanists and particularly Frank Beebe, the enlightened artist, and perhaps one of the last living people who could have written a "Natural History of the World," who drove my life of inquiry.

From the John Green contacts I encountered, it was Bob Titmus who I found wandering the shores and hills of Princess Royal Island that put me into sasquatch fieldwork. I helped him place two large cartons of old Brownie Hawkeye camera traps in good sasquatch habitat.

From the date I released John Green's *The Apes Among Us,* I got an annual visit from René Dahinden, politely threatening me with jail for violating his copyrights as to publishing some of the images Green used that René said he owned. After the threats came discussions of his latest investigations.

Many years later (2005) I republished a book he owned, Roger Patterson's *Do Abominable Snowmen of America Really Exist?* It was republished with a supplement under a different title, and in 2018 I republished just the book itself. This first book and many other Hancock House titles were largely brokered by the region's most prolific and serious sasquatch researcher—Chris Murphy. Thanks Chris.

Science is wonderfully enlightening. If it wasn't bogged down by the individual limitations of scientists, by the taboos and jealousy of professionalism and of the "thematic eagerness" to get grants, science would have long ago overcome human frailties.

This book is another attempt to overcome these human limits. Had it not been for Jane Goodall endorsing this work I would have said it was another scholarly attempt to give credibility to a topic so needing of a "paradigm shift" of the establishment to give it the credibility it deserves! Jane has enabled scientists to make one of the last and greatest paradigm shifts—that animals can display human traits and emotions. Galilei died under house arrest and Bruno at the stake for pointing out the obvious—things that were simply a contradiction to the "science of the day." Meteorites went from an "earth-based science" (of course at a time when everything was earth-centered) to a field of astronomy in another paradigm shift—perhaps some other shifts would today be enlightening. Dmitri Bayanov is perhaps another prophetic researcher of our times.

Enjoy. Be enlightened. Question and don't get hung up on scientists' adoration of themselves; being stuck to last months flavor or their non-scientific tendencies in covering their butts. Remember that fecal analysis is an excellent, accepted ecological tool. Using every tool to follow and analyze one's own trail could be most enlightening. Allow a shift when deserved.

David Hancock
Publisher & Biologist
"Long may the Eagles Fly"

PS: I never put a transmitter on an eagle that I didn't wonder what habitat or creatures it might be so familiar with as it traversed the unknown mountains and valleys of the wild northwest. DH

FOREWORD BY DR. JEFF MELDRUM

In *The Making of Hominology,* senior Russian homin inves-tigator Dmitri Bayanov offers a timely retrospective and introspective consideration of the conception, gestation, diffi-cult labor and imminent birth of a new scientific discipline. In science, names and definitions of terminology are integral to effective communication of knowledge. "Hominology" is the study of "homins," a generic term coined by Bayanov to include all forms of "hairy bipedal primates, whose degree of kinship with humans *(Homo sapiens)* is still to be estab-lished." It is a name essentially synonymous with "relict hominoids," a term first popularized by Boris Porshnev, and carried on, after much deliberation, in the title of The Relict Hominoid Inquiry (www.isu.edu/rhi), a singular academic journal established in 2012, as an attendant to the birth of this revolutionary discipline.

The account of this travail contains numerous dichotomies, highlighting contrasting perspectives, interpreta-tions, politics and paradigms. There have been and continue to be interesting distinctions in the US vs. Soviet institutional approaches to this scientific enigma. There have been and con-tinue to be polarized opinions about the nature of homins— more human-like vs. more ape-like. There have been and con-tinue to be disparities in opinion regarding the uniformity vs. diversity of homins – e.g., Sanderson's Neo-Giants (sasquatch or bigfoot) vs. Neanderthaloids (almas). Exploring these dif-ferences makes for intellectual "dramatic tension" that can breathe vigor into the nascent discipline, desirably, and lend resolution and delineation. To this end, *The Making of Hominology* offers a seminal contribution to the conversation.

With the spirit of the Bolsheviks, but hardly speaking in the majority, Bayanov parallels Kuhn's principles of scientific revolution with the struggles for scientific recognition of hominology, in the midst of a generational paradigm shift. This paradigm shift has turned from the single-species hypothesis, which posited that two culture-bearing hominins could not exist at the same time, an assumption of the com-petitive exclusion principle. Now, however, the fossil record

shows that myriad now-extinct hominins existed simultaneously across the same landscapes. There was not a single-file
line of evolution, but a bushy tree, making room for the possibility that Bigfoot and other relict hominoids could exist. The
question remains, what evidence points to the probability of
such species existing today? Bayanov offers a frank indict-
ment of the scientific communities on both continents, as ones
"duped by the mass media," rendered largely ignorant of the
primary data, in spite of Sisyphean efforts by some to present
the evidence through scholarly channels, and engage objective
discussion. "The theory is the tool that allows you to see the
facts," anthropologist Esteban Sarmiento says. "For people to
see something totally new, they'd need a theory that would
allow for it. Unless the academics have new theories, some
facts will always be closed to them." Bayanov credits
Porshnev as among the first to provide a theory to accommo-
date the "anomalous" facts.

Against this backdrop, Bayanov considers the nature of
some of the most compelling anomalous evidence at present—
the footprints and the film. Smithsonian primatologist John
Napier, one of the few scientists to offer a relatively objective
assessment of the evidence, concluded on the basis of the footprints, that sasquatch does exist. "There must be something in
north-west America that needs explaining, and that something
leaves man-like footprints." (Napier, 1973, p. 205). Bayanov
reprints a poster presentation I delivered at the annual meetings of the American Association of Physical Anthropologists
in 1999. This evaluation commenced in 1996 when I person-
ally examined a fresh track line in the ground, comprised of
15-inch hominoid footprints. More than 20 years later, I have
assembled over 300 specimens of footprints attributed to various relict hominoids around the world. Examples of these are
presented in Chris Murphy's contributed chapter.

It is timely, on the heels of its 50th anniversary, to consider the most compelling photographic evidence—the Patterson
and Gimlin film, taken in 1967. Russian investigators made
significant contributions to the study and analysis of this film.
Here again another dichotomy is revealing—the contrasting
opinions, after regarding the very same film clip, arrived at by

the academicians vs. those of the non-academic professionals, who, it is noted, had no "axe to grind." The history of the alternate perceptions and pronouncements about the film is very revealing of the sometimes glacial-pace of the realization of paradigm shifts in science. To illustrate, consider once again Dr. Napier, who was among the first scientists to critically examine the film in the USA. In his book, he ultimately came to the conclusion that the film was a hoax, although he acknowledged he couldn't put his finger on exactly what to base that conclusion upon. Subsequently, he offered this caveat, "The upper half of the body bears some resemblance to an ape and the lower half is typically human. It is almost impossible to conceive that such structural hybrids could exist in nature. One half of the animal must be artificial." (Napier, 1973, p. 91). In essence, the film subject did not fit commonly held preconceptions of what a hominid should look like, not to mention that the prevailing paradigm would not even allow for the existence of another extant bipedal hominid. Shortly after the publication of his book, a more complete fossilized skeleton a specimen of *Australopithecus afarensis,* among our earliest hominin ancestors (popularly dubbed "Lucy") was discovered in East Africa, and publically announced to much fanfare. The experts were cited in the press as noting how interesting a specimen it was—*from the waist up it looks much like a chimpanzee, while from the waist down it resembled a human.* It seems such hybrids of structure were no longer inconceivable after all. Perhaps the potential of other bipedal homins existing alongside *Homo sapiens* should not be assumed as inconceivable either.

Kuhn has suggested that it may take the passing of a generation before a novel paradigm can take root and flourish. Bayanov, in essence, is issuing a call to action, which if realized, will likely first be responded to by the upcoming generation of scholars. I have seen signs of such germinations and suspect *The Making of Hominology* may contribute to their nurturing.

Jeff Meldrum, PhD
Professor of Anatomy & Anthropology
Idaho State University

INTRODUCTION

As I have taken part in the making of hominology, it's in place here to relate briefly how I came to it and have kept at it, without grants and funding, since 1964 to the present (December 2018); that is for 54 years. I quote from an interview, mentioning some facts of my biography.

As a boy, I was very interested in animals, and visits to the zoo excited me much more than visits to the cinema (not any longer, for I hate seeing animals in cages).

On June 22, 1941, Nazi Germany attacked the Soviet Union, and history's most terrible war started. Moscow was repeatedly bombed, and my father took the family (mother, sister, and me) to Tajikistan (then part of the Soviet Union), far away from the front. My father was soon called-up into the army, and my mother, sister and I stayed in Tajikistan until the end of the war in 1945, when we returned to Moscow.

While in Tajikistan, we lived in a small town called Shakhrinav and it was there and then that I heard for the first time a rumor of "hairy wild men" living in the mountains; but could hardly believe this. I recalled it decades later when I revisited Tajikistan in 1982 on a hominological expedition, as described in one of my books.

At the time, I continued to entertain great interest in animals and dreamed to become a famous zoologist; like the eminent German naturalist Alfred Brehm (1829–84), /... /

As a youth and young man, back in Moscow, I gave much thought to what happened to mankind as a result of a second and much nastier world war—about what happened to the German people under Hitler, and the Soviet people under Stalin. From zoology my interests shifted to philosophy, sociology, and anthropology. It was due to these interests that in 1964 I met Professor Boris Porshnev, who acquainted me with the problem of so-called relict hominoids, and that was the start of my hominological investigation, including the Bigfoot phenomenon. (*Bigfoot Research: The Russian Vision*, pp. 389–391).

My first book on the subject (in Russian) was titled *Wood Goblin dubbed "Ape": A Comparative Study In Demonology*. I wrote this book after expeditions to the Caucasus; interviewing witnesses and reading a great many books on folklore and demonology. My conclusions: such folklore terms as devils, goblins, brownies, etc., mean real biological beings for the local population in many geographical areas where such terms are traditionally used.

With great amazement, I had discovered a giant gap between the tenet of science and what common locals take for the truth. For instance, Professor of psychology Constantin Platonov cites the words of an old hunter in Siberia, who told him, "I don't know if apes exist or just imagined, but I saw the Leshy with my own eyes, and more than once," as an example of belief in sheer superstition. Why? Because the Russian-English dictionary translates "Leshy" as "wood-goblin," and "goblin of the woods," with the designation "folklore." It's appropriate to add that an eyewitness in Siberia claims to have encountered a "hornless devil."

In this respect, it is instructive to quote from Theodore Roosevelt's book *Wilderness Hunter: Outdoor Pastimes of an American Hunter* (1893), wherein he recounts an episode in the life of a hunter and trapper, named Bauman, who was hunting and trapping with a partner in the mountains of Idaho /Montana in the early 1800s. His partner discovered near their camp some tracks of a bear that "has been walking on two legs." Bauman laughed at this ... At midnight he was awakened by some noise, his nostrils were struck by a strong, wild-beast odor; he caught the loom of a great body in the darkness ... Eventually, when the two men parted for a time, Bauman's partner was killed by the beast-like stranger. After that "Bauman, utterly unnerved, and believing that the creature

15

with which he had to deal was something either **half human** or **half devil,** some great **goblin-beast,** abandoned everything but his rifle and struck off at speed down the pass ..." (Quoted from Ivan T. Sanderson, *Abominable Snowmen: Legend Come To Life,* 2006, pp.105–108, my emphasis – DB).

So it's not accidental that the nature of such "devils" and "goblins" is at present thoroughly investigated by diverse scholars whose findings are posted by the Relict Hominoid Inquiry at Idaho State University.

After my book on folklore and demonology about hairy wildmen in Eurasia, I wrote a similar work of 69 pages regarding such denizens of North America, titled *"Learning from Folklore" (Russian Hominology,* 2016, pp. 43–112). I based it mainly on the material in the book *Giants, Cannibals & Monsters: Bigfoot in Native Culture,* 2008, by anthropologist Kathy Moskowitz Strain. My conclusions were similar or identical with those of my first book: Native Americans using various ethnic names (regarded in North America as mythological), view Bigfoot (or Sasquatch) as biologically real beings. When I asked Kathy about my conclusion, she answered, "I think Native people view Bigfoot as a relative—but the kind you don't really want to invite to Christmas dinner."

Was it important and relevant to the making of hominology? Very much so. The first reason used by the angry academic critics and opponents of Porshnev—accusing him of creating and spreading pseudoscience—was his alleged presenting of non-existent mythological creatures as being real. So where is the truth in this crucial issue: on the side of native people around the world or on the side of the scientific community? This question was the first chain that tied me to this research. The second was different but no less strong, which I now relate.

As a young man, going on expeditions, I sort of polled the natives on their views and experiences. Getting older, staying more at home, dealing more with the theory than practice of hominology, I started to poll educated people around me on their knowledge of themselves, so to speak. Namely, what they thought about the species name of ourselves—*Homo sapiens*. It's Latin; in English it means literally "wise man." Strange, isn't it?

> ... man, proud man,
> Drest in a little brief authority,
> Most ignorant of what he's most assured,
> His glassy essence, like an angry ape,
> Plays such fantastic tricks before high heaven
> As make the angels weep. (*Measure for Measure*, II, ii)

Can such creature be called *Homo sapiens?* So I kept asking my friends, acquaintances and others: Who coined the term *"Homo sapiens,"* when and why? I don't remember anyone answering it correctly. What separates man from animals, including non-human primates, is human language. Therefore I put my question recently to Noam Chomsky, who is:

> ...an American linguist, philosopher, cognitive scientist, historian, political activist, and social critic. Sometimes described as the "father of modern linguistics," Chomsky is also a major figure in analytic philosophy and one of the founders of the field of cognitive science." (Wikipedia)

His answer received in November 2018 was: "I'm one of those who doesn't know who or when [as to the *Homo sapiens* term], though I think it's clear why. It's a very distinct species, and 'sapiens' is at least a hopeful designation."

Actually, if educated humans, including anthropologists, philosophers and linguists do not know the origin of their own species name, this is a "designation" that the current paradigm in anthropology is in error and inadequate. To use John

Napier's words, "*Homo sapiens* is not the one and only living product of the hominid line." Why then was this crucial fact not known to science earlier? Because there was no science to know it. The Darwinian revolution is still incomplete and going on. As a result, the needed science was born as late as the second half of the 20th century and as described now in *The Making of Hominology: A Science Whose Time Has Come*, now in print by Hancock House Publishers.

That its time has come is shown, for example, by the work "The Patterson-Gimlin Film in Light of the Linnaeus and Porshnev Teachings" 7:97-101 (2018), posted by the Relict Hominoid Inquiry (RHI) On-line Journal (website); making quite explicit the origin of the term *Homo sapiens*. Also by the work "The Ecology of an Uncatalogued Hominoid of the Boreal Forest (Taiga) of North America and Eurasia," by Dr. John Bindernagel; posted on the Sasquatch Canada website and shedding a bright light on the ecological issue of hominology.

What's more and very important is that hominology has reached the DNA testing level in its development, as testified by such works on the RHI as "DNA as Evidence for the Existence of Relict Hominoids," 5:8-31 (2016) and "Normal Science, Revolutionary Science: Notes on Bryan Sykes' *The Nature of the Beast,*" 4:75-78 (2015)

Finally, Dr. Henry Bauer, in his mind-opening book *Science Is Not What You Think (*2017) advocates establishment of a Science Court whose sole mission "would have to be limited strictly to clarifying purely scientific issues about which there is dispute" (p. 209).

The Making of Hominology is a call to the scientific community— to primatologists, anthropologists and paleoanthropologists—to abandon assumptions turned into erroneous convictions; a call to heed the on-going revolution in the science of man and shift the paradigm. Dr. Jane Goodall is describing the situation as "one of the greatest unsolved mysteries in the natural world." In fact, the "mystery" remains largely a mystery because facts and arguments shedding light on it have been stubbornly hushed up and ignored by the dominant mainstream scientists. Hopefully, this book will help put an end to the impasse even before the establishment of a Science Court.

Dmitri Bayanov
Moscow, Russia
December 3, 2018

Links for Referenced Material

The Patterson-Gimlin Film in Light of the Linnaeus and Porshnev Teachings 7:97-101 (2018):
<https://www.isu.edu/media/libraries/rhi/from-the-editor/Bayanov_-PGF_50th.pdf>.

The Ecology of an Uncatalogued Hominoid of the Boreal Forest (Taiga) of North America and Eurasia:
<https://www.sasquatchcanada.com/uploads/9/4/5/1/945132/paper_for_website_main_-_bindernagel_pdf.pdf>.

DNA as Evidence for the Existence of Relict Hominoids:
<https://www.isu.edu/media/libraries/rhi/research-papers/HART-DNA-Evidence.pdf>.

Normal Science, Revolutionary Science: Notes on Bryan Sykes' *The Nature of the Beast* 4:75-78 (2015):
<https://www.isu.edu/media/libraries/rhi/comments-amp-responses/Commentary-on-Sykes.pdf>.

CHAPTER ONE
SETTING THE STAGE

Quote of the year: "All I can say is that it has been at least 67 years and we are still ESSENTIALLY at square one" (Chris Murphy).*

This striking metaphor, referring to the history and present status of hominology, deserves thorough scrutiny and discussion. Its truth is in the fact that like 67 years ago, so at present, the official dictum of anthropology is that *Homo sapiens* is the only living representative of the genus *Homo* on earth. Any information about different extant hominids (hominins) is the result of hoaxes, fabrications or mistakes. In this respect, yes, we are still at square one. The truth of this situation is most deplorable. Yet there is another great truth: the fact that our "square one" is quite different today from what it was 67 years ago or, say 20 years ago, when Russians and North Americans (Grover Krantz and John Green) celebrated together the 30th Anniversary of the Patterson and Gimlin film at the Darwin Museum in Moscow. At the time there was no International Center of Hominology in Russia or the Relict Hominoid Inquiry in the US.

So let us look at the history of our research and try to understand why despite so many expeditions, books, articles, forums, conferences, exhibits, and other events and endeavors "we are still ESSENTIALLY at square one." It all began with the Yeti Himalayan expeditions in the 1950s, but further on the

*Chris included all of the 1950s in this figure. There was research being done during this time and even earlier, but it did not get "serious" until 1957, so 60 years would be more appropriate.

main players and most activity were concentrated in Russia (the Soviet Union at the time) and in North America. Ivan Sanderson (1911–1973) stood out in America, Boris Porshnev (1905–1972) in Russia. The former as author of *Abominable Snowmen: Legend Come To Life,* 1961; the latter as author of *The Present State of the Question Of Relict Hominoids,* 1963. The Russians came up unexpectedly with a most ambitious and scholarly project. The Soviet Academy of Sciences formed a special Commission to investigate the "snowman" question. It was organized at the initiative of Porshnev; academician Sergey Obruchev became its head, Porshnev was his deputy. The Commission began collecting and studying historical and current information on the subject and publishing it. What's more, the Academy launched a big expedition in 1958 to search for the snowman in the Pamir Mountains.

This is what Sanderson wrote in his book about the events in Russia:

These Soviet activities shed an entirely new light on the whole business, and also put it on such an altogether higher plane that Western scientific circles were obliged to change their attitude toward the matter quite drastically (p.19).

Further on, Sanderson explains:

This was that the whole problem is an anthropological rather than a zoological matter. In other words, all the Sino-Soviet evidence pointed to the ABSM [Abominable Snowmen] being primitive *Hominids* (i.e., Men) rather than *Pongids* (i.e., Apes) or other nonhuman creatures..." (p. 20).

This important distinction continued always to play a major role in this research. Referring to "Western scientific circles," Sanderson wrote that:

21

No longer could they simply avoid the issue by saying that it had not been explained or that its protagonists were merely a bunch of amateur enthusiasts pursuing a fantasy.

Mentioning the discovery of the coelacanth fish, he remarked:

This had at first been called a hoax, but finally had to be accepted as living proof of the fact that not everything about the life of this planet is known. Obviously, creatures confidently thought to have been decently extinct for tens of millions of years can still be around (p. 20).

Porshnev's volume came to light in 1963 in an edition of only 180 copies because by that time the attitude of the Soviet Academy toward the snowman problem had radically changed. Porshnev had many opponents in the Academy and they won the day when the Pamir expedition returned empty-handed. Snowman Commission head Sergey Obruchev did not approve of Porshnev's idea of relict Neanderthals. He said that he would have never been involved in the snowman problem had he thought they were Neanderthals. He believed it was an unknown bipedal ape. Porshnev said the opposite: "I would never have been involved in the problem of the snowman had I thought it was an *ape.*" Note how strongly and decisively the dichotomy and opposition of human and non-human primates divided the researchers. As a result, the Snowman Commission was dissolved. Fortunately, some time later its activities were resumed by the members of the Smolin Relict Hominoid Seminar at the State Darwin Museum, but without any academic or state support and involvement. On the academic level the theme and subject became virtually illegitimate, dubbed pseudoscience, and tabooed.

It would require considerable text for me to describe the contents and import of Porshnev's 1963 great volume, existing only in Russian. It was great both in size and contents, embracing and summarizing diverse evidence dug up and collected by the Snowman Commission during a couple of years of its existence. I will only focus on some points concerning the birth of hominology. Finding a suitable name for the objects of our research was a great problem showing how unprepared science was for its novel task. The moniker "abominable snowman" was too ridiculous to be used in science. Sanderson could find nothing better than using the acronym ABSM throughout his big volume. In Russia the word "abominable" was dropped and the name "snowman" alone was widely used, which was not a scientific name either. The term "relict hominoid" in the very title of Porshnev's book had been suggested by Pyotr Smolin, chief curator of the Museum, and accepted by Porshnev. Both used it in an etymological, not taxonomic sense, meaning a "relict humanlike primate." They wanted to distinguish it from the term "bipedal anthropoid," used by those who believed the snowman was an ape. Porshnev wrote of the two versions explaining the nature of snowmen: "anthropoid" and "hominid." To use "hominid" for the snowman in a direct taxonomic sense was too bold at the time, so "hominoid" was chosen. Still Porshnev was bold enough to equate the snowman, and consequently, relict hominoid, with Neanderthal, using also "paleoanthrop" as a synonym. That was possible because he believed that Neanderthals had no speech, no language, and therefore could not be classed fully human. He must have been mistaken in his hypothesis, but I am not going to discuss this moot question here. What I accept at present is his view that the origin of speech and language is the main factor in the origin of man (language—the "rubicon of mind"—St. George Mivart), and that the problem is still facing us. Secondly, I accept the idea that the "wildmen" of Europe, especially its western part, are relict Neanderthals.

What was new in the book and absent in the works of scholars in the West is Porshnev's resurrection of Carl Linnaeus's pri-

23

mary role in our research and his contribution to it. He devoted two pages to Linnaeus in the beginning of the book and, what's most significant, Porshnev baptized the snowman, alias relict hominoid, alias relict Neanderthal and paleoanthrop, by the scientific name of *Homo troglodytes Linnaeus*. Thus Chapter 11 is titled "Preliminary description of *Homo troglodytes* L. ('snowman'). Morphology."

In his documentary story *The Struggle for Troglodytes* (*La Lute Pour Les Troglodytes* – Heuvelmans, 1974*), Porshnev devotes several pages to Linnaeus and to my work on the subject. I'm pleased to report that Paul LeBlond has now translated this material from French to English and it is posted on the RHI website <http://www2.isu.edu/rhi/pdf/PORSHNEV-FORMATTED.pdf >. As Porshnev's student and follower, I also give credit to Linnaeus in my four books in English. In one of them his name is present on 12 pages, in another on 26. Historically, it is Linnaeus who gives justification to the subtitles of modern books, such as *Legend Come To Life* and *Legend Meets Science.*

Another unique feature of Porshnev's book is his mention in it of "the arising science of the relict hominoid" (p. 273). This is most important. He viewed our research as a new scientific discipline. To my knowledge, this fact is not yet brought home and appreciated by our colleagues outside Russia, in North America in particular. After Porshnev's passing in 1972, when we studied and verified the Patterson and Gimlin film, I felt very strongly the need for a proper and fitting name of the science which we relied on in our work. It was then that I coined the word *hominology.* To some degree the term has already taken root in the literature of our research, both in Russia and abroad. Unfortunately, hominology is often viewed as an integral part of cryptozoology which is wrong. In fact, cryptozoology is not a scientific discipline, but a "Scotland Yard" of zoology, with a special kind of methodology of searching for cryptic, i.e., very elusive, animals. Plants are not elusive, so there is no cryptobotany. When a cryptid is discovered and identi-

*Heuvelmans, Bernard; Boris F. Porshnev (1974). *L'homme de Néanderthal est toujours vivant*. Paris: Plon.

fied (e.g., gorilla, okapi, giant squid) it is turned over to a corresponding discipline of zoology. Thus there is no special discipline of cryptozoology.

Hominology, on the contrary, IS a special discipline, and very specific at that, just because its subject-matter is very specific and special. Why then is it ascribed to and confused with cryptozoology? This is because its objects of study are most elusive and skillful in avoiding those who are looking for them. So hominology and cryptozoology share a similar or even identical methodology of searching for their objects of interest. Still one is part of zoology, the other of anthropology. It can be thus said that hominology is still in a *cryptoanthropological* stage of development. Lack of clarity in this matter has always been detrimental to hominology which I'll touch upon later.

Thus the words *hominology* and *hominologists* were indispensable for the birth of the new discipline. What about the terms *relict hominoid, relict hominid* or *hominin?* Their simultaneous use is caused by our lack of exact knowledge regarding the taxonomic level of the "uncatalogued" primates in question, as well as by changes of taxonomy in primatology itself. Regarding the latter, Thomas M. Greiner, Associate Professor of Anatomy and Physical Anthropology, writes the following:

When scientists use the word hominin today, they mean pretty much the same thing as when they used the word hominid twenty years ago. When these scientists use the word hominid, they mean pretty much the same thing as when they used the word hominoid twenty years ago. /.../ If you're more confused now than you were before, you are just about where you should be. We scientists really need to clean up shop in this area" (Thomas M. Greiner, "What's the difference between hominin and hominid?")

"Cleaning up shop" in hominology, I felt a need for a handy single word to indicate the objects of our research. For a preliminary working term my final choice stopped at *homin*. Hominology is the science of homins, hairy bipedal primates, whose degree of kinship with humans *(Homo sapiens)* is still to be established. It's expected to be different for different species or subspecies of homins. Homin is a synonym of such terms as relict hominoid, hominid and hominin, also used when need be. So the problem of terminology, so bad when we started up, is no longer acute and crippling.

CHAPTER TWO

SCIENTIFIC LOOK AT SCIENCE

If hominology claims to be a science, while the Establishment rejects the claim and calls it a "pseudoscience," this contradiction calls for a scientific look at science itself and scientific literacy in particular. In this connection I asked a lot of educated people why their scientific name is *Homo sapiens* and who coined this glorious name. I never got a correct answer.

So what is science? Modern science is such a vast enterprise and activity that it can't be defined by one sentence. Several are needed to embrace all of its sides and aspects which I don't need now. Mindful of our theme and subject, I chose the following: **"Science is the pursuit and application of knowledge and understanding of the natural and social world following a systematic methodology based on evidence."** Here is another: "Science is a logical method to answer questions about the natural universe." I like here the word "logical" because science is based on and guided by logic. But the most crucial word is missing in these definitions. It is present here: Science is "a never ending search for truth" (Ann Druyan). Yes, truth is science's *conditio sine qua non*. So here is my definition: Science is humanity's main Hunter for Truth.

Then what is truth? **"Truth means being in accord with fact** and **reality."** What is not science? Religion and mythology are definitely not. A dictionary definition of myth: "an idea or story that is believed by many people but that is not true." If mythology is not true, how can it be a database for hominology? And if science is a search for truth, is science always right? Do scientists make mistakes? Yes, they do. There is a readable book on scientific blunders committed by great scientists. Mistakes happen in science. Science isn't perfect.

> Science will never give full understanding of anything. All that we can hope for are useful approximations of the objective reality we hope is out there. A common claim about the superiority of science over other ways of knowing is that science is self-correcting; science may take wrong turns from time to time, but it eventually finds its way back on the right road. /.../ However, it's important to understand how human limitations—scientists are human, after all—sometimes undermine the process of self-correction ("Self-correction in science" Posted by Surak). Ultimately, science is only as dependable as the humans who apply it. (Ronald A. Lindsay)

Science has a long history of development; today it is different from what it was in the distant past or even not so long ago. The process of scientific development is best described in one book, "which may be the most influential treatise ever written on how science does (or does not) proceed" (John Horgan). That book is *The Structure of Scientific Revolutions* (first edition 1962, second edition 1970) by Thomas Samuel Kuhn (1922–1996), an American physicist, historian and philosopher of science. For Kuhn, the problem was twofold: (i) to explain why scientific theories are accepted, and (ii) to explain why scientific theories are replaced. These two aspects are intimately related, and the key concept that Kuhn develops is that of "paradigm"—a reigning or dominant approach to solving problems in a given area of science. (Thomas Kuhn's "Theory of Scientific Revolutions")

The Structure of Scientific Revolutions was influential in both academic and popular circles, introducing the term *paradigm shift,* which has since become an English-language idiom. Kuhn made several notable claims concerning the progress of scientific knowledge: that scientific fields undergo periodic paradigm shifts rather than solely progressing in a linear and continuous way, and that these paradigm shifts open up new approaches to understanding what scientists would never have considered valid before; and that the notion of scientific truth, at any given moment, cannot be established solely by objective criteria but is defined by a consensus of a scientific community. /.../ Thus, our comprehension of science can never rely wholly upon "objectivity" alone. Science must account for subjective perspectives as well, since all objective conclusions are ultimately founded upon the subjective conditioning/worldview of its researchers and participants.

Kuhn challenged the prevailing view of progress in "normal science." Normal scientific progress was viewed as "development-by-accumulation" of accepted facts and theories. Kuhn argued for an episodic model in which periods of such conceptual continuity in normal science were interrupted by periods of revolutionary science. The discovery of "anomalies" during revolutions in science leads to new paradigms. "New paradigms then ask new questions of old data, and move beyond the mere 'puzzle-solving' of the previous paradigm" (Wikipedia)

The new paradigm is popularized in text-books, which serve as the instruction material for the next generation of scientists, who are brought up with the idea that the paradigm—once new and revolutionary—is just the way things are done. The novelty of the scientific revolution recedes and disappears, until the process is begun anew with another anomaly-crisis-paradigm shift" (Thomas Kuhn's Theory of Scientific Revolutions).
<http://bertie.ccsu.edu/naturesci/PhilSci/Kuhn.html>

CHAPTER THREE

COMMENTS ON KUHN'S THEORY AND EXAMPLE OF ITS APPLICATION

To quote Thomas Kuhn:

> Normal science, the activity in which most scientists inevitably spend almost all their time, is predicated on the assumption that the scientific community knows what the world is like. Much of the success of the enterprise derives from the community's willingness to defend that assumption, if necessary at considerable cost.
>
> Normal science, for example, often suppresses fundamental novelties because they are necessarily subversive of its basic commitments (p. 5).

These concepts are supported by the practice and history of science, as well as by the writings of other scientists.

> Thus, human beings, including scientists, do not function under continual awareness of humanity's fundamental ignorance; rather, they live under perpetual illusion of fundamental understanding (Henry Bauer, *Scientific Literacy and the Myth of the Scientific Method*, 1994, p. 75).
>
> It is a self-satisfied dogmatism that cherishes the delusion that our available knowledge is somehow infallible and final. However, this contradicts the very essence of science, in which the most exciting and significant discoveries lead to dramatic changes in

world view (Kuhn 1970). Nonetheless, reception of novelty in science is typically analogous to the reaction incurred with foreign tissue in a host: rejection (Beverly Rubik, *Life at the Edge of Science,* Hemi-Sync Journal, Vol. XII, No. 2, Spring 1994).

So one can comprehend that innovators in science routinely encounter resistance if their ideas are sufficiently original; almost invariably, if those ideas contradict significant parts of the conventional wisdom; the more strongly, if novel methods are also involved; and, of course, more emphatically if the innovator is relatively unknown or an outsider to the relevant specialty (Henry Bauer, Ibidem, p. 75).

Do you have any idea how vicious the scientific establishment—guys like you—can be when something threatening comes along? It's like any bureaucracy with a vested interest in the status quo, only worse. If a new theory surfaces that contradicts accepted wisdom, it's shot down—bang!—as soon as it's picked up on the radar. God forbid it should penetrate and get through to the masses!
    If it's only threatening, it's subjected to ridicule. Journals weigh in, academics scoff, the popular press writes funny stories. But if it's something revolutionary like this, they play hardball and it gets the full treatment. Careers are ruined, people are run out of town, nothing appears in print. No one wants to look foolish. /.../ The point of the story is the way the establishment reacted—the way it always reacts. It prefers to blot out something for which it has no ready explanation. It's an old story, older than Galileo. Science will turn to superstition and torture to defend its right to be wrong" (John Darnton, science-fiction novel *Neanderthal,* 1996, p. 57–inspired by Boris Porshnev's idea of relict Neanderthals and hominology activities in Russia).

John Darnton was editor for *The New York Times* and must have known scientists "with a vested interest in the status quo." It is they, not science, who "defend their right to be wrong."

So scientific truth and new paradigms are won not in events like long-distance marathon running, but more like a steeplechase, i.e., an athletic obstacle race. For illustration, let me repeat an example I cited 40 years ago in the book *The Scientist Looks at the Sasquatch*. I borrowed it from *Encyclopaedia Britannica* and it began like this: "History of Meteoritics: Since very ancient times men have known about meteorites falling; however, the scientific study of meteorites is hardly older than 150 years." Today one can get much information about the history of this discipline, which is a branch of astronomy, especially from the book *Cosmic Debris: Meteorites in History* by John G. Burke (University of California Press, 1986). On one hand, ancient and anecdotal information on "stones from the sky" was connected with myths and legends, which raised a red flag for the scientists in the age of Enlightenment. On the other hand, from Aristotle to Newton astronomers seemed to believe that the heavens could not be littered with stones. So in the 18th century, members of the French Academy of Sciences, then the highest authority in all scientific matters, "were convinced that such an irregular phenomenon as the fall of a stone from heaven was impossible, and preferred to doubt all the reports of witnesses and to change their statements to conform with acknowledged scientific theories." They had several to explain away the anomaly without a change of paradigms. Shooting stars were seen as "an atmospheric phenomenon, like lightning, or they were explained by atmospheric processes, such as showers of hail condensing in clouds." "Stones from heaven" were also explained as "terrestrial rocks that had been struck by lightning"—hence the name "thunderstones." Others believed that meteorites were volcanic rocks, violently spewed out during major eruptions.

A change in paradigms was on its way in the last decades of the 18th century. In 1772, during one of his travels through the remote areas of Siberia on behalf of Czarina Catharina, the renowned German naturalist Peter Pallas examined a huge iron mass near the town of Krasnojarsk—a mass that the Tartars said had fallen from the sky. The 700 kg iron mass caught the scientist's attention—it was partly covered with a black crust, and there were many translucent olivine crystals (peridots) set in its iron matrix, something Pallas had never seen nor heard about. Unwittingly, he had discovered a new type of meteorite, a class of stony-iron meteorites that would later be named for him: the pallasites.

Pallas' subsequent report encouraged a German physicist, Ernst Florens Chladni, to publish his audacious thesis that this and other finds actually represent genuine rocks from space. In his booklet, *On the Origin of the Pallas Iron and Other Similar to it, and on Some Associated Natural Phenomena*, published in 1794, he compiled all available data on several meteorite finds and falls. From this, he was forced to conclude that meteorites were actually responsible for the phenomena known as fireballs, and more importantly, that they must have their origins in outer space. His view received immediate resistance and mockery by the scientific community. In the late 1790s, rocks from space just didn't fit into the concept of nature. However, nature itself came to Chladni's aid in the form of two witnessed meteorite falls, making him the father of a brand new discipline—the science of meteoritics./.../

Nevertheless, a large number of conservative scientists kept on denying the obvious facts, among them some of the most influential members of the respected French Academy of Sciences. Their mockery and sarcasm was silenced several months after a publication by British chemist Edward Howard on his analysis of a 25 kg stone. Then on April 26, 1803, a shower of about 3,000 stones fell in broad daylight near L'Aigle, France, witnessed by countless people. This incident attracted much public attention, providing a fertile ground for further research and the young science of meteoritics. The French Minister of the Interior commissioned the young physicist Jean-Baptise Biot, a member of the French Academy of Sciences, to investigate the fall, resulting in a

well-written paper that finally broke the spell. The L'Aigle fall and Biot's subsequent publication caused a scientific landslide; a change in paradigms that had been prepared in time by Chladni and Howard, establishing beyond any doubt the fact that meteorites are genuine rocks from space.
(Meteorite.fr - Basics - History of Meteoritics—)
<http://www.meteorite.fr/en/basics/meteoritics.htm>

What a remarkable example of Thomas Kuhn's explanation of how science proceeds! Following it, my conclusion 40 years ago was this: "Since hominoids cannot be expected to fall from the blue on their own, as is the case with meteorites, I wish some pranky UFOnaut would dump a load of bigfeet on the heads of skeptics among modern academics" (*The Scientist Looks at the Sasquatch*, Edited by Roderick Sprague and Grover S. Krantz, The University Press of Idaho, Moscow, 1977, p. 155).

Since my North American colleagues did not learn obvious lessons from this example, I'll try to point them out now. At the time of Aristotle, astronomy was, to use Kuhn's dictum, a "normal science." At the time of Galileo Galilei and Giordano Bruno it became revolutionary. As a result, Galilei ended his days under house arrest and Bruno was burnt alive at the stake. At the time of Newton, astronomy became normal again and astronomers again knew "what the world is like," especially the sky, which is their field of knowledge. Suddenly, to their surprise and resentment, steps in a physicist (Ernst Chladni) and declares that he takes it for fact that some stones happen to drop on earth from the skies. His view receives "immediate resistance and mockery." Astronomers say all witnesses are liars or superstitious people. All samples, presented as stones from outer space, can be explained by their earthly origin. So what is the difference between the physicist and the astronomers? It is in the fact that his statement is based on knowledge, theirs on ignorance. And here I dare offer my own contribution to the famous Kuhn theory. **A paradigm shift is necessarily accompanied and followed by a corre-**

sponding expertise shift. Isn't it instructive that, although meteoritics is a branch of astronomy, its originator Ernst Chladni was not an astronomer? The old paradigm astronomers were no longer taken for reliable experts on the issue. Yes, science is self-correcting and building upon itself, but the most important correcting is not done by those who erred.

The supporters and followers of the new paradigm started building a new discipline, creating first of all its terminology and a classification system. The Greek word for "meteor" meant just "thing in the air" (hence similarly sounding words "meteoritics" and "meteorology"). A specialist in meteoritics is named a meteoriticist. A meteorite is a meteor that reaches earth's surface; those composed of metal are called siderites; of stone aerolites; of metal and stone siderolites. The discipline is a normal science today. In the past 200 years, it has matured to a highly interdisciplinary field, with its own mineralogical and radiometric methods of investigation. Say meteoriticists proudly: "Today, we have unlocked the secrets of the origin of some of the meteorites in our collections. Isn't it great that we recognize samples of the far side of the Moon, the surface of the planet Mars, and several asteroids in our collections—without having to invest in very expensive space missions first?"

Also of educational value for hominologists is this information:

The Meteoritical Society is a non-profit scholarly organization founded in 1933 to promote research and education in planetary science with emphasis on studies of meteorites and other extraterrestrial materials, including samples from space missions that further our understanding of the origin and history of the solar system.

The membership of the society boasts over 1,000 scientists and amateur enthusiasts from over 40 countries who are interested in a wide range of planetary science. Members' interests include meteorites, cosmic dust, asteroids and comets, natural satellites, planets, impacts, and the origins of the Solar System.

CHAPTER FOUR

HOMINOLOGY IN THE LIGHT OF KUHN'S THEORY

In March 1966, the Soviet journal *Questions of Philosophy* printed under the rubric "For discussion" Professor Boris Porshnev's article "Is a scientific revolution in primatology possible today?" This is its very beginning in my translation:

> If we use Thomas Kuhn's concepts (T. Kuhn, *The Structure of Scientific Revolutions*, Chicago, 1962), then in "normal science" in the stage of "anomaly" and "crisis," new evidence, incompatible with the existing paradigm, is not only hushed up, but also censured and ridiculed. Though it does not give pleasure, it's high time to state that in the science of man's origin there accumulated certain anomalies and that attempts to explain them within the existing paradigm proved ineffective. According to T. Kuhn, this means an impending "scientific revolution."

Though this was only four years after Thomas Kuhn enriched science with his theory and the term *paradigm shift* had not become yet an English-language idiom, Porshnev was already taking account of Kuhn's ideas in developing his own theory. This is most significant. This means that almost from the beginning Russian hominologists have been aware of Kuhn's explanation of science's *modus operandi* and tried to act accordingly. This was not the case with our overseas colleagues. Thomas Kuhn's name and ideas appeared in North American hominology only 44 years later, in John Bindernagel's book *The Discovery of the Sasquatch,* 2010, and even then only in relation to "normal science."

Regrettably, the idea of "revolutionary science" is never mentioned or taken into account by our North American colleagues in connection with the Bigfoot/Sasquatch research and phenomenon.

The scientific importance of the latter was duly marked and stressed by John Green, expressing surprise at the attitude of the scientific community in this case. For example, he wrote:

> This is not a game or a fantasy; it is a question of serious scientific research of tremendous importance. It may not have the glamour of moon shots, but in what it can teach us about our origins and our physical potential it may be even more important (*The Sasquatch File,* p. 70).

> The fascination of the subject, for me, involves the very thing that I am most inclined to complain about— that the scientific world ignores it. The material available should be of great interest to more than one branch of science (Sasquatch, 1978, p. 11).

> ...of the billions of research dollars and millions of man and woman hours of scientific talent, hardly a dollar or an hour is devoted to this quest. Why that should be so is, to me, the most intriguing mystery of all (*Journal of Scientific Exploration*, Vol. 18, No. 1, 2004).

What was for Green "the most intriguing mystery of all" was not a mystery at all for Porshnev and his followers in Russia familiar with the history of science and its operation as described by Thomas Kuhn. For the benefit of North American readers and researchers, John Green reproduced in his books in translation a fair amount of Boris Porsnev's material, but he never paid attention to the fact that Porshnev was not just investigating, but preparing a new scientific discipline and a "revolution in science." His above-mentioned article ends with these words:

> We can see that the scientific revolution under discussion means not just a discovery of new facts or a collection of them. It means a radical transformation of a whole scientific discipline, but such a transformation that doesn't reject even a grain of earlier obtained factual knowledge. The need is only in changing the glasses.

The article, offered for discussion, was sort of covered up, as no discussion followed, which was unprecedented in the history of that respectable academic journal. So now I turn to another figure who played a role in the making of hominology, the figure of an outstanding British primatologist and paleoanthropologist, Dr. John Napier, who made a contribution of sorts to our research by his book *Bigfoot: The Yeti and Sasquatch in Myth and Reality,* 1973. I did a more or less detailed review of that book in Chapters 5 and 6 of *America's Bigfoot: Fact, Not Fiction*, 1997, calling Napier's book "a protracted attempt to explain Bigfoot away" and "this masterpiece of evasiveness and equivocation." Decades later I wrote that John Napier, being a typical worker of normal science, "tried to blacklist all evidence of our revolutionary science, including the Patterson/Gimlin documentary film." But what I want to note here is not my critique, but some of Napier's insights and excellent generalizations, deserving to be repeated and remembered. In fact, without using the terms "paradigm shift" and "revolutionary science," he spelled out what these notions really mean in relation to anthropology:

> But if any one of them [Bigfoot footprints – DB] is real then as scientists we have a lot to explain. Among other things we shall have to re-write the story of human evolution. We shall have to accept that *Homo*

sapiens is not the one and only living product of the hominid line, and we shall have to admit that there are still major mysteries to be solved in a world we thought we knew so well (*Bigfoot*, 1973, p. 204).

No one doubts that some of the footprints are hoaxes and that some eyewitnesses are lying, but if one track and one report is true-bill, then myth must be chucked out of the window and reality admitted through the front door (p. 203).

One is forced to conclude that a man-like form of gigantic proportions is living at the present time in the wild areas of the north-western United States and British Columbia. If I have given the impression that this conclusion is—to me—profoundly disturbing, then I have made my point. That such a creature should be alive and kicking in our midst, unrecognized and unclassifiable, is a profound blow to the credibility of modern anthropology (Preface to the 1976 edition of *Bigfoot*).

CHAPTER FIVE

THEORY, METHOD AND PROOFS

B ut if it's something revolutionary like this, they play hardball and it gets full treatment" (John Darnton). Boris
Porshnev did proclaim and advance something revolutionary
and therefore could not escape "hardball and full treatment"
from the opponents. In 1969, four Soviet zoologists, three of
them professors and one academician, attacked Porshnev with
their collectively authored article "Pseudoscience in the Guise
of Searchers for Neanderthaloids." They protested that
Porshnev's monograph *The Present State of the Question of
Relict Hominoids*, with a circulation of only 180 copies, was
allowed to be published at all. They called it "This unprecedented shameless similitude of a scientific compilation ..."
and mentioned the Patterson and Gimlin film in this way:

> In northern California, some loafers made a film of a
> Bigfoot, a wild monster with dangling breasts which
> was fabricated on the spot out of a boy or girl dressed
> up in a bear skin, while our magazine *Znanie – Sila*,
> No. 9, 1968, reported with enthusiasm this balder-
> dash!

The article also had such passages:

> As for a "scientific" organization of the search for the
> "snowman," it has now become completely clear that
> there is more purpose in first studying the organizers
> and participants of such a search in a physiological
> laboratory on higher nervous activity. /.../ All the above
> makes us ask the question: Do the people circulating

such yarns have the right to bear the honorary title of a Soviet scientific worker? (*Vestnik Zoologii*, 1969, No. 4, pp. 69–80).

No wonder Porshnev was voted down in the election for the membership of the prestigious Soviet Academy of Sciences. He was refused the possibility to answer his critics directly and this stimulated him to write a documentary story *The Struggle for Troglodytes* in which he shed light on his side of the issue (as noted earlier on page 24 the work has been translated by Paul LeBlond). But no publisher dared to publish it in Moscow. It was printed in the city of Alma-Ata, Kazakhstan, in the literary magazine *Prostor*.

Also, for many years, he had worked on what he regarded as his *magnum opus*: *On the Beginning of Human History: The Problem of Paleopsychology*. Getting permission for its publication had taken him a lot of time for it demanded decisions of various commissions. Permission was given, but at the last moment the editor took fright and publication was banned. Shortly after the author had a fatal heart attack. This happened in November 1972. Porshnev's friends and colleagues strongly protested the outrage, and the book, minus some parts, was published in 1974. A full edition saw light in 2007.

Not long before his passing, Porshnev shared with me his apprehension that with his possible death (he said he did not expect to live long), our research and activity would come to an end. With a strong feeling, I tried to assure him that we, his students and followers, would do our best to carry on the cause and work of his creation. I am happy the promise proved to be true. Many of Porshnev's ideas, especially in the book *On the Beginning of Human History,* will have to be tested and verified when homins are studied in depth and in earnest. But what he had the time to work out and we picked up and developed is the very theoretical basis of the discipline we call now hominology. This theory is responsible for the appearance and existence of hominology's paradigm.

As said earlier, science is based on and guided by logic. That is by reason and common sense. How come? We see the Sun moving and the Earth standing still. Yet, science tells us that it is the Earth that is moving and going around the Sun. Is that logical? Yes, it is, and became so as soon as a new paradigm, that of heliocentrism, took the place of geocentrism in the mind of Copernicus, Galileo and those who followed them. To think that the Earth is the motionless center of the world, as was believed earlier, is quite illogical and irrational, for it means that the whole universe turns in 24 hours, at unimaginable speed, round our planet. Logic, i.e., correct reasoning and clear thinking, tell us that it is the Earth's rotation on its axis that explains the visible movement of the sky. The point here is that there is no other proof of the Earth's movement than those provided by logic, reason and common sense, even though our sense of vision testifies to the contrary. Logic and reason are the attributes of intellect and philosophy, while our senses appeared and developed in evolution not for the sake of philosophy, but for the sake of survival of our animal ancestors. Sense organs work well for that mundane purpose, but they are not fully reliable in the "heavenly" work of science and philosophy in the quest for truth.

The Copernican revolution in science won victory long ago; the Darwinian, which hominology is part and follow-up of, is still in progress. And again, there is no other proof of Darwinism and evolution than those provided by logic and reason. As objects of hominology are relict beings, it stands to reason that humans have coexisted with them throughout history and therefore we must have both historical and modern evidence of relict hominids. Inevitably, they must have been known to scholars of antiquity and the Middle Ages. That is exactly what we learn. Here is a fine example from antiquity:

According to Plutarch, when the Roman general Sulla (old spelling Sylla), having sacked Athens in 86 BC, was returning to Italy, he came to Dyrrachium (modern Durres in Albania):

In this place, we are told, a satyr was taken asleep, exactly such as statuaries and painters represent to us. He was brought to Sylla, and interrogated in many languages who he was; but he uttered nothing intelligible; his accent being harsh and inarticulate, something between the neighing of a horse and the bleating of a goat. Sylla was shocked with his appearance and ordered him to be taken out of his presence (Plutarch, 1792, 349).

Homins' presence in medieval Europe is well documented in the book *Wild Men in the Middle Ages*, 1952, by Richard Bernheimer:

About the wild man's habitat and manner of life, medieval authorities are articulate and communicative. It was agreed that he shunned human contact, settling, if possible, in the most remote and inaccessible parts of the forest, and making his bed in crevices, caves, or the deep shadow of overhanging branches. (p. 9)

Remember *Homo troglodytes L*, forest (sylvestris) and cave-dweller:

Medieval writers are fond of the story which tells how hunters, venturing father than usual into unknown parts of the forest, would chance upon the wild man's den and stir him up; and how, astounded at the human semblance of the beast, they would exert themselves to capture it, and would drag it to the local castle as a curiosity ... (p.17)

In many ways his life resembled that which we now attribute to the raw beginnings of human cultural existence in the **Stone Age.** (p.10) (My emphasis – DB)

It must be said that the author's views and logic are paradoxical; despite the striking realism of the evidence presented in the book, he takes pains to assure the reader that the hero of the book—wild man—is an imaginary creature! This is a result of the author's absolute adherence to the mainstream in science; in other words, his adherence to normal science.

Hominology is rich in mythological types and images: satyrs, goblins, devils, shaitans ... A sure sign, say the opponents, that hominology is based on mythology, not reality. On the contrary, says the hominologist, if homins had nothing to do with folklore and mythology that would be a sign of their non-reality. For folklore and mythology ARE based on reality. If images of real animals are present in folklore and mythology, how could the creators of myths and legends fail to pay attention to the "wild men?"

The same logic demands the presence of homin images in ancient and medieval art, and sure enough, such images are there worldwide and on a large scale. They have certain specific features, hairiness in the first place, which help tell them apart from human images.

As noted, hominology still exists in a cryptoanthropological stage of development. That is to say hominologists need and seek all signs of homins presence in nature, their footprints in particular, which is a necessary and logical evidence of the beings reality. The science of tracks—ichnology—is readily used in zoology and paleontology, as well as by Criminal Investigation Departments. Homin footprints have certain differences from footprints of *Homo sapiens*, and this is also a necessary and logical quality of this type of evidence.

If homins are real and present on all the inhabited continents, they can't escape being encountered and sighted by humans. This is the fifth most abundant kind of evidence in the hands of hominologists. These five categories of evidence are, at the same time, the initial basic stock of hominology's database and proof of homins' existence:

1. Natural history
2. Folklore and mythology

44

3. Ancient and medieval art
4. Footprint evidence
5. Eyewitness testimony

It is impossible to prove the existence of meteorites to those who are convinced that stones cannot fall from the sky. As a closer example, Rudolf Virchow, the foremost 19th century German anthropologist, member of the Prussian Academy of Sciences, declared that the Neanderthal skull, found at the time, was that of a deformed human. Virchow was a vehement anti-evolutionist. As more and more hominid fossils were unearthed, a dogma set in, invented by paleoanthropologists—that of a total extinction of pre-sapiens hominids. This was and remains a gross misconception, based on nothing but ignorance. Hominologists have no chance to prove anything to those who lack knowledge in this matter and refuse to learn. That is until the misconception is overturned and done away with.

Each scientific discipline, moreover a revolutionary one, has its own special proofs and methodology, in accordance with its own paradigm. Here are thoughts on methodology by reputable scholars. Scientific method is:

Nothing but trained and organized common sense (Thomas Huxley).

All sorts of method have combined to allow and to foster the growth and progress of science (Henry Bauer).

It is all too frequently supposed that scientific method was discovered, and once discovered, that was that. It was then there to be used, and change in science has resulted from the regular use of the tool... Nothing could be further from the truth. We continually make discoveries in science, and there is every reason to suppose that we make discoveries in the area of methodology as well (William Newton-Smith,

Canadian philosopher of science, author of *The Rationality of Science*).

And here is a novelty of methodology introduced by the father of hominology:

> It is precisely the use of nontraditional methods, such as the comparative analysis of mutually independent evidence that has made it possible to establish the existence of this relic species and to describe its morphology, biogeography, ecology, and behavior. In other words, fact-finding methods have been used in biology that are usually employed by historians, jurists, and sociologists (Boris Porshnev).

Here's another remark of his on proofs and method in hominology:

> There is no need to demand that the neck vertebrae of Louis XVI be put on the table to prove that he was guillotined... This fact is accepted as proven by another, not less scientific method.

Finally, the brand of pseudoscience applied to hominology by the establishment is absolutely unjustified. According to Henry Bauer, the difference between science and pseudoscience is in the presence or lack of what he terms as "connectedness," i.e., "crucial links with the mainstream." Modern science, he writes, is "strikingly coherent across its various disciplines." "Interdisciplinary projects are commonplace." (*Science or Pseudoscience*, 2001, pp. 11, 158). I understand this as follows. The unknown can only be studied and understood by proceeding from and connecting with the known. By this criterion, UFOlogy is not yet a science because so far UFO observation reports cannot be connected with or explained by the existing scientific knowledge.

Hominology, on the contrary, by the criterion of connectedness seems to be the most scientific of sciences, for it provides "crucial links" with and between the theory of evolution, paleoanthropology, mythology, demonology, folkloristics, the history of religion, and the history of art.

In addition, hominology gives a natural answer to the natural question why apes are still with us while brainier apemen or pre-sapiens hominids died out. The answer is they didn't. Their whole extinction is the illusion of paleoanthropologists who are as adequate experts on relict hominids as paleontologists were on living coelacanths. Relict hominids are hidden in natural forests and mountains, but above all they are hidden in "the forests of the mind." The task of hominology is to drive them out of those "forests" into the open vistas of science (*Bigfoot Research: The Russian Vision*, p. 144).

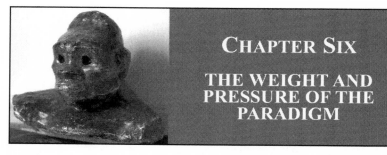

CHAPTER SIX

THE WEIGHT AND PRESSURE OF THE PARADIGM

The sixth category of the hominology database is Photographic Evidence, still represented only by the glorious Patterson and Gimlin film (PG film). It's still standing alone because of its superior quality in comparison to numerous other photos and videos whose lack of clarity and detail excludes them from scientific analysis. This is in accord with our knowledge of the great problem of getting homin photographic evidence. We owe the clear images of Sasquatch/Bigfoot on film to Roger Patterson's courage and adroitness. Deborah Martyr once sighted Orang Pendek at a distance of 30 meters. Says she: "I had a camera in my hand at the time, but I dropped it; I was so shocked!"

The PG film, whose 50th anniversary we marked in 2017, plays a special role in the making of hominology. The history of its presentation, rejection, analysis and verification went on in perfect agreement with Thomas Kuhn's theory of scientific *modus operandi* in the face of scientific anomaly and crisis. For an appraisal and study of his film Roger Patterson applied to different institutions and the highest among them was the Smithsonian Institution, whose reaction was similar to that of the French Academy in the 18th century in the case of meteorite falls. Regarding the Patterson and Gimlin film the Institution's verdict was this: "The recent movie film taken in northwestern California cannot be regarded as scientific proof as there is no evidence that the creature seen in the film is all it has been claimed to be." A more explicit response was given by Dr. Richard W. Thorington, Jr., Director, Primate Biology Program, on November 26, 1971. He began with the explanation "...why many of the Smithsonian scientists considered the film to be a joke. To them it appeared all too obvious that the pictures were made of a person dressed up in an ape costume, trying to run in an unnatural way." As a matter of fact, there is no Sasquatch running or trying to run in the film. Further on Dr. Thorington explained his attitude to the "Bigfoot phenomenon":

> One should demand a clear demonstration that there is such a thing as a Bigfoot before spending any time on the subject. There are many, many valid areas of research for which the subject matter is known to exist, so one should busy oneself with these rather than with will-of-the-wisps.

History was made by the judgment of Dr. William Montagna, "the distinguished director of the Regional Primate Center at Beaverton, Oregon":

> Along with some colleagues, I had the dubious distinction of being among the first scientists to view this few-second-long bit of foolishness. As I sat watching the hazy outlines of a big, hairy man-ape taking long, deliberate human strides, I blushed for those scientists who spent unconscionable amounts of time analyzing the dynamics and angulation of the gait and the shape of the animal, only to conclude (cautiously, mind you!) that they could not decide what it was! For weal or woe, I am neither modest about my scientific adroitness nor cautious about my convictions. Stated simply, Patterson and friends perpetrated a hoax. As the gait, erect body, and swing of the arms attest, their Sasquatch was a large man in a poorly made monkey suit. Even a schoolchild would not be taken in.

According to John Green, "On the basis of reports from their Seattle representative, *Life* magazine editors showed some interest, until they had the film run through for some people at the American Museum of Natural History—who took one quick look and wrote it off as a fake."

Next is the historic trip to Europe and Russia in 1971 by the Canadian Sasquatch and fortune hunter René Dahinden whose pros and cons in hominology are described in detail in the book *America's Bigfoot: Fact, Not Fiction*. This is how the aim of his trip is described in the book:

> Was he to believe American science that Sasquatch was nothing but an Indian legend and that plaster casts of giant footprints in his possession and the film itself were after all nothing but fraudulent? Had he been chasing a wild goose? Fed up with scientists in America he decided to seek the answers from science in Europe, particularly in England and Russia. In England because she was the *Alma Mater* of the Yeti searches in the Himalayas which set him on the Bigfoot track in the first place, and in Russia because he had heard of Professor Porshnev and colleagues in Moscow (pp. 26, 27).

In England René left a copy of the PG film for analysis by Dr. John Napier who then presented his findings and judgment in the book *Bigfoot: the Yeti and Sasquatch in Myth and Reality*. I quote the highlights of Napier's analysis regarding the film:

> If we confine ourselves rigidly to what most scientists would regard as hard evidence, then the answer is heard loud and clear: *Bigfoot does not exist...* There are no skulls, no postcranial bones, no captive animals, no photographs or cine films of unquestionable probity. What possible justification is there for intelligent people to countenance such a wraith? (p. 197).

Incidentally, Napier dubs Bigfoot fans and researchers "monster enthusiasts" and "monster establishment." He wrote:

> The best item of the California Bigfoot saga I have kept to the last—the Roger Patterson film. In theory

this should have wrapped up the whole thing. One can hardly quarrel with a movie taken at the range of approximately 100 feet showing a continuous sequence 20 feet long of 16 mm colour film of a Sasquatch head on, in profile and in rear-view; but of course it all depends on the movie—and the star (p. 89).

Napier's view of the film subject's anatomy:

The upper half of the body bears some resemblance to an ape and the lower half is typically human. It is almost impossible to conceive that such structural hybrids could exist in nature. One half of the animal must be artificial. In view of the walk, it can only be the upper half (p. 91).

As we see in Roger Patterson's film, the Sasquatch is acquiring a somewhat gorilla-like image superimposed on a basically human framework. This turn of events had done nothing for the credibility of the legend (p. 203).

There I think we must leave the matter. Perhaps it was a man dressed up in a monkey-skin; if so it was a brilliantly executed hoax and the unknown perpetrator will take his place with the great hoaxers of the world. Perhaps it was the first film of a new type of hominid, quite unknown to science, in which case Roger Patterson deserves to rank with Dubois, the discoverer of *Pithecanthropus erectus* or Java man; or with Raymond Dart of Johannesburg, the man who introduced the world to its immediate pre-human ancestor, *Australopithecus africanus* (p. 95).

Conclusion:

There is little doubt that the scientific evidence taken

collectively points to a hoax of some kind. The creature shown in the film does not stand up well to functional analysis. There are too many inconsistencies (p. 95).

In Moscow, Russia, Dahinden gave public film showings and lectures at a number of scientific organizations and media editorial offices. At the Institute of Ethnography reaction to the PG film of cultural and physical anthropologists was negative:

Some argued that there could not possibly be such creatures in North America. Some said that they could not survive in such habitats, that they should not be hairy, that they can't have hair-covered breasts, all the usual things that American and Canadian "specialists" had been saying to Dahinden. (*America's Bigfoot: Fact, Not Fiction* (p. 30).

At the Institute and Museum of Anthropology René had a long and friendly talk with the "patriarch of Russian anthropology" Mikhail Urisson, who was also a friend of Porshnev, Koffmann, Burtsev* and myself. The film was not shown to Urisson then but when he saw it later, he said to us:

I wish you success, but if you want my advice, then never show this film around and never refer to it. It can compromise everything else. Great apes never lived in North America, and therefore this film cannot be true.

To Dahinden's surprise, there was one research institution at which the film showing was a complete success: the Central Scientific Research Institute of Prosthetics and Artificial Limb Construction. In Dahinden's words, "the whole joint came to a standstill. Most of these people returned to work after one

*Name was changed from Bourtsev to Burtsev after publication of this material and subsequent material.

52

screening of the film, while six of 'the top people' remained and watched the film several times." Dahinden sensed that something important was happening as the scientists, speaking in Russian, intently discussed the movements of the filmed Bigfoot. He asked what they were saying and the comments were translated to him: "What we are seeing here is something extremely heavy, in combination with something extremely bulky."

Dahinden asked the scientists how they knew the creature was heavy. They explained that great weight was indicated by how the creature's arms were swinging and how the knees were bending as the weight of the body came onto the feet. The knees of the creature were bending to absorb tremendous weight. Dahinden states:

> That's when I realized I made a mistake by going and talk-ing to anthropologists and zoologists. They didn't see what was there. You don't go to a physical anthropologist or a biologist to show them a Sasquatch film if there is any question about authenticity. These people in the artificial limb outfit, they had no axe to grind. They really didn't give a damn if the Sasquatch exists or not. You know, they did-n't have a position to defend like an anthropologist or zool-ogist. They were looking at the locomotion and the move-ment, and that rather impressed me.

As a result of that showing, the director of the Institute, Professor N. I. Kondrashin, wrote a letter on January 6, 1972, to the chairman of the USSR Committee on Cinematography, A.V. Romanov, answering a request about the credibility of the Patterson and Gimlin film. The letter stated:

> The film was viewed and discussed by our Institute's spe-cialists in man's locomotion. The film contains sufficiently clear frames of the walk of a manlike creature, a detailed study of which would undoubtedly be of serious scientific interest. (*America's Bigfoot: Fact, Not Fiction* (pp. 31, 32).

Now isn't this a perfect lesson? Doesn't the above speak loudly and clearly about the weight and pressure of this or that paradigm on the mood and mentality; nay, the very sense of vision of corresponding specialists? And about the reality of the expertise shift in case of the paradigm shift?

René Dahinden left us a copy of the PG film and copies of footprint casts in January 1972, and at the end of that year Bayanov and Burtsev informed him and other colleagues in North America that the film *was real and the subject shown in it was NOT a man in a suit.* How could that be? Neither Bayanov nor Burtsev was a primatologist, anthropologist or paleoanthropologist. How could they be right and scores of PhD specialists in America and Russia be wrong? Yes, they *could* because there had already happened a *paradigm and expertise shift* in that field of research. All specialists who followed the old paradigm in primatology and anthropology could not be experts in the new science of relict hominoids based on the theory proclaimed by Boris Porshnev. As his student and follower I recognized and fully accepted the existence of relict hominids as proved by the evidence of the five database categories mentioned earlier. In this respect our attitude to the PG film from the start was quite different from the attitude of analysts who denied the existence of relict hominids or were skeptical. With a new paradigm, accepted by us, eventual appearance of a Bigfoot movie documentary appeared quite normal and expectable, in a sense even necessary.

To use Napier's language, "the monster enthusiasts" had "no photographs or cine films of unquestionable probity." Could we really prove that the Patterson and Gimlin film was true and real? Was it worth the time and effort? Yes, it certainly was. A film is the most visual kind of evidence after a live specimen, and vision is one of the most important senses in man. If the documentary film was true and what it showed was real, it appeared as valuable as our other kinds of evidence combined. The question was whether we, young hominologists, would be able to perform the task of verification, because some old researchers had passed away and others, including Porshnev and Smolin, were in poor health.

The director of the Research Institute of Prosthetics and Artificial Limb Construction advised the Chairman of the Committee on Cinematography to study the Patterson and Gimlin film, thinking that it was of "serious scientific interest." The copy of the film we had was left by Dahinden to Igor and my opinion was that he should not hand over the copy to the Cinematography Committee but work on it himself. Igor agreed and undertook a tremendous job of producing usual size photographs from the tiny frames of the film. That was essential for a thorough and productive analysis of the footage. Igor and I combined our talents and knowledge in different disciplines in our work on the film and the result turned out positive. Besides photography, of special value was Igor's use of mathematics in determining the subject's height and weight, as well as the speed of the camera's shooting. But that was found later and was not present in our first report to colleagues in North America. It should be noted that accuracy of Igor's calculations depended on the measurements supplied by Dahinden, as obtained by him on the film site. Unexpectedly, Igor and his wife Alexandra, inspired by the film, revealed great talent at sculpting and produced fine scale models of what we called Patty Bigfoot (so named after Patterson).

Now about our first report of film analysis. From the very beginning we realized that we were not sufficiently equipped and enlightened to extract all possible or necessary information from the filmstrip. So our aim was to determine whether the filmed subject was different from modern man and in agreement with our idea of a relict bipedal primate. For that purpose we widely resorted to paleoanthropological information and comparisons with the results of Dr. Krantz's work on Sasquatch footprints. We used the terminology employed at the time by Russian anthropologists. The title of our five-page report was this: "Preliminary Notes on the Materials of American Hominologists: The Patterson 1967 Footage, Photographs and Plaster Casts of Footprints, Professor Krantz's Paper 'Anatomy of the Sasquatch Foot.'"

Here is a somewhat shortened text with some editing:

GENERAL REMARKS ON THE FILM – Roger Patterson's filmstrip shows a hairy man-like creature, walking erect, having well-developed breasts and buttocks. The last three points, if we accept for a time the authenticity of the creature, indicate its belonging in the Hominid, not the Pongid (anthropoid), line of evolution of higher primates. Morphology of the head shows a very outstanding brow ridge, a low bridge of the nose, very pronounced prognatism, and a cone-shaped back of the head. Judging by the well-developed breasts the creature is female. However, the muscles of the back, arms and legs are so much in relief that they call for comparison with those of a heavy weightlifter. The creature "has no neck," or at least the neck is not to be detected at first sight. Looking back the creature turns its upper torso along with the head to a much greater extent than would normally a human being. This might indicate a somewhat different attachment of the skull to the spine than in man, and a strong development of the neck muscles which conceal a short, sort of simian, neck. /.../

LOCOMOTION AS SEEN IN THE FILM – It seems smooth and resilient like that of a big quadrupedal animal. One gets the impression that the creature steps on slightly bent legs. If that is the case the impact on the heels should be less manifest than in man's walk, and the hominoid tracks, usually rather even in depth, seem to corroborate this conclusion. While walking the creature swings its arms intensely using them as walking beams as it were.

COMPARISON TO SUPPOSED GAIT OF NEAN-DERTHALER – Prof. Boris Porshnev, who put forward the Neanderthal hypothesis on the relict hominoid origin, in his monograph (1963), page 288, refers to the opinions of Russian anthropologists V.P. Yakimov, G.A. Bonch-Osmolovsky and V.V. Bounak concerning the walk of Neanderthalers as construed by analysts of fossil material. We find it very significant that the two characteristics mentioned above—i.e., less impact on the heels and arms swinging—are listed by anthropologists as supposed traits of Neanderthal locomotion, while slightly bent legs are ascribed to Neanderthalers even in a standing position.

THE HOMINOID FOOT – The main features standing out in both the American and Soviet (Russian) material: 1. Tracks show flat feet (without an arch). 2. The width of the foot in proportion to the length is much greater than in man's foot. 3. The hominoid foot is generally much bigger than man's.

Besides, as has been often noted by late Pyotr Smolin, chairman of the Hominoid Problem Seminar at the Darwin Museum in Moscow, the hominoid foot is distinguished by a great mobility of its toes which can bend very much or fully extend or spread very widely.

One more peculiarity: the so-called double ball at the back of the big toe as evidenced in many North American tracks (Green, 1968; Krantz, 1972). We find Grover Krantz's explanation of this feature very interesting, and we especially value at this stage the conclusion drawn by him concerning the size of the creature's calcaneus (heel bone). In some frames the creature's foot seems to have an unnaturally protruding heel. To a casual observer this may look like a sticking-out edge of an artificial sole, but to those who know better this is an omen of the creature's reality.

As for the double ball itself we would like to make here the following remark. The double ball is made up not only of two bulges of tissue but also of a furrow between them, which is like a kind of fold on the sole. Hence the question can also be put this way. Why is a fold formed at this spot on the hominoid sole?

The answer, probably, can be like this: because the hominoid foot is not so rigid as man's foot, it still retains a certain measure of mobility inherited from the hand-like foot of the ape, and therefore has a furrow somewhat analogous to lines on man's palm.

Grover Krantz finds the correlation between the great weight of the creatures in question (as evidence, among other things, by the depth of footprints) and the anatomy of the foot, as it is revealed in the very same footprints, so natural and binding that he makes the following conclusion: "Even if none of the hundreds of sightings had ever occurred, we would still be forced to conclude that a giant bipedal primate does indeed inhabit the forests of the Pacific Northwest."

It's the first time such an unambiguous statement is made by a professional anthropologist regarding the problem of relict hominoids, a statement made even more welcome by the fact that it came about as a result of study of material evidence which is the plaster casts and photographs of footprints.

COMPARISON TO THE NEANDERTHAL FOOT – As far as we know, none of the American researchers has compared the hominoid foot, as revealed in footprints, to the Neanderthal foot, reconstructed on the basis of fossil material.

In the Soviet Union this job has been done by Prof. B.F. Porshnev who noted a similarity in such features as lack of an arch, the width to length ratio, great mobility of toes (Porshnev, 1963).

It seems that a new and very important development in this direction of research is a comparison made by us between the calcaneus (heel bone) of the Neanderthal foot and that of North American Hominoids as shown in the materials of American hominologists. Grover Krantz, on his part, concludes that the Bigfoot has "enlarged heels," so "the heel section must be correspondingly longer." He also writes that the creature's "ankle joint must be set relatively farther forward along the length of the foot," so its length is expected to be "set relatively farther forward on the foot than in man."

Thus, this is also true of the Neanderthal foot, dramatically illustrating the above point.

To make things even more fascinating, the very same features show on the foot of the creature in Roger Patterson's filmstrip. To our knowledge, this fact has not been mentioned before by analysts of the film.

It follows that in analyzing a possible anatomy of the hominoid foot we find agreement in three, apparently independent sources: 1. Roger Patterson's film; 2. Photographs and plaster casts of footprints obtained by René Dahinden and others, and analyzed by Grover Krantz; 3. Morphology of the Neanderthal foot.

NEANDERTHAL OR PITHECANTHROPUS? – /.../ As for the giant size of North American hominoids, we think this cannot be a sufficient argument against Porshnev's standpoint since big variations in size are also true of the species *Homo sapiens.* Yet, there is, in our opinion, one serious obstacle to identifying the Patterson film creature with a relic of the Neanderthal stage of evolution, which is that the creature's head is too much ape-like. (That was said BEFORE Igor discovered a frame of the film showing that Patty's face was more human-like than ape-like! - DB) /.../) If that is so, we can expect that in certain areas of the earth there remain relict "Neanderthal beasts," in other areas – "Pithecanthropus beasts," still in a third – mixed forms of the former two or even other forms. It is believed that the evolution of the family *Hominidae* (or *Troglodytidae,* in Porshnev's classification) proceeded at such a fast pace (in terms of evolution) that the forms it created were, so to speak, on the move and genetically open, not set and sealed like species created in a very long and slow evolutionary process.

In our analysis we did not refer to *Gigantopithecus* because virtually nothing is known about that form of primates except their giant size. As for what is known of the foot of *Australopithecus* and *Homo habilis,* it does not seem to fit the pattern of the hominoid foot we are dealing with.

NOT MAN-MADE – So our conclusion at this stage is the following: though it is not yet clear in what relation North American hominoids stand to the making of man, it is pretty clear now they themselves are not man-made.

Moscow, October 1972. Dmitri Bayanov, Igor Burtsev

As I re-read this report, written nearly 45 years ago, it struck me by its grip on the issue and difference of approach and attitude from those of the people like Montagna, Thorington and Napier. Our objective and open-minded analysis had not revealed even a hint of a hoax in the Bigfoot documentary. We pointed out that the filmed specimen was not an ape, nor a *Gigantopithecus,* but considering all indications and characters, a *relict hominid.* In short, in contrast to critics and opponents, we acted as representatives of a novel discipline, named right then *hominology.*

The report was published in 1973 by Don Hunter with René Dahinden in the book *Sasquatch,* and in 1975 by Peter Byrne in *The Search for Big Foot.* A shortened version is now posted on the Internet: The First Russian Report on the 1967 Bigfoot Film <http://www.bigfoot-lives.com/html/the_first_russian_report_on_th.html>.

The next important steps in the issue were Igor's discovery of a frame in the film footage showing Patty's face far more distinctly than in other frames, and Alexandra's "discovery" of a specialist in biomechanics, Dr. Dmitri Donskoy. Igor's discovery had done away with Napier's judgment that the filmed Sasquatch had the head of an ape and the legs of a man, and gave a strong impulse to our further analysis. As to biomechanics, we had certainly noticed peculiarities and differences in Patty's gait and movements, and looked forward to getting a specialist's analysis of them. Some months later Dr. Donskoy provided his report and it was also published in the books by Hunter with Dahinden and by Peter Byrne, along with our report. John Napier ignored the latter (Bayanov and Burtsev were small fry for him), but strongly objected to the opinion of Dr. Donskoy: "The report leaves an inevitable impression that Dr. Donskoy is the possessor of a closed mind; he believes in the Sasquatch and he won't hear a word against it." My comment on Napier's was this:

> What a brazen example of laying one's own fault at somebody else's door! Almost every word of this charge is the opposite of the truth. Dr. Donskoy not only did not believe in the Sasquatch while examining the film, but he never even used the word in his analysis. He had hardly heard of the Sasquatch problem before and is not keen on it now. What aroused his interest, when he was shown the film, was the spectacle of a walk which he found spontaneous and at the same time unbelievably different from the usual pattern. Being a specialist in biomechanics, whose business it is to study and classify human and animal movements, he was amazed and intrigued by that spectacle and found it deserving his time and thought for closer scrutiny (*America's Bigfoot, Fact, Not Fiction*), pp. 78, 152).

Dr. Donskoy's position was similar to that of the specialists at the Institute of Prosthetics and Artificial Limb

Construction. René Dahinden said about them: "they had no axe to grind." Here are concluding sentences of Dr. Donskoy's report:

> Since the creature is man-like and bipedal its walk resembles in principle the gait of modern man, but all its movements indicate that its weight is much greater, its muscles especially much stronger, and the walk swifter than that of man. /.../ On the whole the most important thing is the consistency of all the above-mentioned characteristics. They not only simply occur, but interact in many ways. And all these factors taken together allow us to evaluate the walk of the creature as a natural movement, without any signs of artfulness which would appear in intentional imitations. At the same time, with all the diversity of human gaits, such a walk as demonstrated by the creature in the film is absolutely non-typical of man.

Thanks to Porshnev's high international reputation, one of his papers, in my translation, was published in the December 1974 issue of *Current Anthropology*, a world journal of the science of man, whose founder and editor was a very progressive and open-minded anthropologist, Sol Tax. Porshnev was no longer with us and publication was done on the condition that his views expressed in the paper would be explained and defended by his follower(s). The task was undertaken by me, helped somewhat by Igor (I asked him to undersign the text as a co-author for certain logistic and administrative reasons). I managed not only to present there Porshnev's original views on the evolution of higher primates and answer the critics, but also mention most important hominological data, including Sasquatch footprints and the 1967 Bigfoot documentary:

> The Patterson film, which at last makes the creature's photographic appearance and movements available to everybody's eyes. /.../ ...we have studied the pho-

tographs and plaster casts of footprints ascribed to relict hominoids, on the one hand, and the Patterson film, on the other. /.../ We have established five solid correlations between the footprints and the creature seen walking in the Patterson film, all five distinct from or totally nonexistent in *sapiens* characteristics. This leaves no doubt in our minds whatsoever that both the film and the footprints we studied are genuine (*Current Anthropology*, Vol. 15, No. 4, December 1974, pp. 454, 455).

The next *Current Anthropology* Volume (CA 16: 486-87) carried Sasquatch researcher Gordon Strasenburgh's comment on my article where he claimed that *Australopithecus africanus (Paranthropus)* was the ancestor of relict hominoids. This resulted in publication of my next paper in *Current Anthropology,* titled "On Neanderthal vs. *Paranthropus*" with the following headings of its parts: The Uniqueness of Hominids; The Riddle of Neanderthal Disappearance; Some Ancient and Medieval Evidence; Eighteenth-century Evidence; Evidence from the Caucasus and Central Asia; A Quick Look at the American Hominoids. The article was well illustrated with ancient and medieval homin images and mentioned my negative view of the *Gigantopithecus* ancestor hypothesis (CA June 1976).

To my knowledge, that was the last top penetration by the revolutionary science of hominology into the academic domain of normal anthropology. My next contact with *Current Anthropology* happened many years later, in 2002, when I offered the journal's new Editor, Benjamin S. Orlove, an article for publication. His reply was polite but firm: "I appreciate your interest in the journal. However, your manuscript does not fit the scope of the journal, therefore I am unable to accept it for publication."

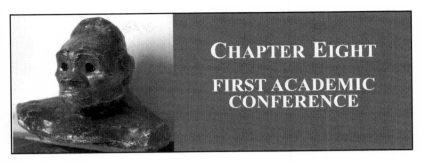

Chapter Eight

FIRST ACADEMIC CONFERENCE

Yes, the "scope" of the journal is different today from what it was back in the 70s, when *Current Anthropology* contributed so nicely to the making of hominology. As it happened, in December 1975, a letter, sent from the University of British Columbia, Vancouver, BC, arrived at the Darwin Museum in Moscow, addressed to myself and Igor Burtsev. It said:

> Cyril Belshaw has shown me your paper, "Neanderthal vs. *Paranthropus*" and I was most excited that the range of phenomena which you consider overlaps significantly with that I hope to explore in a conference to be called "Sasquatch and Other Monsters." A statement about the conference is enclosed.
>
> /.../ I would like to invite you to consider joining with us. If you are interested, and will suggest a paper title and abstract, we will see what is involved in bringing you here.
>
> It is true that in the past North American anthropologists have refused to take Sasquatch and similar phenomena seriously, /.../ **This seems to be an idea or topic whose time has come** (my emphasis – DB). I very much hope that you will be able to share in its exploration with us.
>
> Sincerely,
> Marjorie M. Halpin
> Assistant Professor of Anthropology

In addition to being Assistant Professor of Anthropology, Halpin was also curator of ethnology of the Museum of Anthropology at the University of British Columbia, and one of her next letters to us said the following:

Dear Sirs,

Please accept my congratulations on your article in the latest issue of *Current Anthropology*. Not only do I find your argument convincing, but consider it an important breakthrough in hominology that it appeared in an international journal of this stature.

I am all the more excited about the possibility of your coming to Canada to participate in the proposed conference.

That was how the first ever scientific Sasquatch Conference was conjured up and came about in Vancouver, BC, in May 1978. This historic event is reported in some detail in *America's Bigfoot: Fact, Not Fiction*, and here I need only mention some of its outstanding facts and features. The conference name "Sasquatch and Other Monsters" jarred on my ears, so I suggested another title: "Sasquatch and Similar Phenomena" which was accepted and held for some time, but in the end the organizers added to it "A Conference on Humanoid Monsters."

Our paper title was "Analysis of the Patterson-Gimlin Film and of some Footprints ascribed to the Sasquatch: Why we find them Authentic." It was presented as co-authored by Bayanov, Burtsev and Dahinden (the latter in gratitude for providing us with the film copy and footprint photos and casts). Ours was the one and only report on the PG film and this was one of the most amazing and significant facts of the conference. As to Marjorie Halpin's intention to bring us to Canada, that was not enough for realization without the Soviet authorities' permission to let us go. Permission was refused and our report was presented *in absentia*.

Its impact and significance are indicated by newspaper reports sent to me by Marjorie Halpin. Such as these, for example:

The Russian findings were the hit of a recent academic conference on "humanoid monsters" sponsored by the University of British Columbia in Canada.

Taking obvious delight in the pronouncement, Bayanov said his team's success in authenticating the film was "a triumph of broad-mindedness over narrow-mindedness, which serves a world in need of such success" (*San Francisco Chronicle,* May 31, 1978).

Russian scientists led by Dmitri Bayanov of the Darwin Museum in Moscow are convinced the star of the October 1967 film, shot at Bluff Creek in northern California by Roger Patterson, is indeed a genuine Sasquatch. (*The Vancouver Sun,* May 15, 1978).

The conference organizers promised to publish conference reports in a separate volume. During the year that followed rumors reached us in Moscow that there were problems with selection of papers for publication. At last, we received a letter from Dr. Halpin, dated June 29, 1979, which said the following:

This is a difficult letter for me to write, since in it I must inform you that the decision has been made not to publish your paper in the volume of selected papers presented at the Sasquatch conference.

Excluded were also reports by Marie-Jeanne Koffmann, by well-known anthropologist Dr. Carleton Coon, Dr. Grover Krantz, Dr. Vladimir Markotic, and others.

There was yet another big scandal in the history of hominology. When Dr. Halpin said "an idea whose time has come," she forgot to take into account Thomas Kuhn's theory of scientific revolutions—the tendency of suppressing novelties brought about by new ideas. Fortunately, the "banned" papers, including ours, somewhat shortened and divided, were published in 1984 by Dr. Vladimir Markotic in the book *The Sasquatch and Other Unknown Hominoids,* Western Publishers, Calgary, Alberta. In a review of it, Dr. Roderick Sprague, University of Idaho, mentioned our analysis of the PG film:

> The first section is the second part of the Bayanov, Burtsev, and Dahinden report entitled "Analysis of the Patterson-Gimlin Film: Why we find it authentic." It is by far the best and most thorough discussion of this classic film (*Cryptozoology,* 1986, Vol. 5, p. 105).

Our entire paper is presented in *America's Bigfoot: Fact, Not Fiction,* 1997, pp.107-158. Here is relevant information:

> Glimpses of the Patterson film in various television shows had left me incredulous that the creature shown in it could be real. This book has made me almost equally incredulous that the film could have been faked, and thus I have become open to the staggering possibility that relict hominids may still be with us in sufficient numbers that we have the chance to learn something about them. I recommend this book heartily as a highly interesting reading adventure." (Dr. Henry H. Bauer, Professor Emeritus, Book Review, *Journal of Scientific Exploration,* Vol. 18, Number 3, Fall, 2004, p. 533.)

Of recent origin on hominology's "square one" are DNA analyses by Dr. Melba Ketchum and Dr. Bryan Sykes. An evaluation of their work is given by Dr. Haskell V. Hart in the article "DNA as Evidence for the Existence of Relict Hominoids" – RHI Journal, Research Paper, 5:8-31 (2016).<http://www2.isu.edu/rhi/pdf/HART-DNA-Evidence.pdf>
I'd like to comment on the following extract of the article:

For nearly fifty years following the famous 1967 Patterson-Gimlin Film (PGF) the only evidence for the existence of a relict hominoid (RH), which includes Sasquatch, Bigfoot, yeti, yeren, almasty, yowie, orang pendek, and other "large hairy ape-men" worldwide, was based on eyewitness accounts (numbering in the thousands), footprints, vocalizations, and a very few other videos and pictures, none as convincing as the PGF. Even the PGF itself is still debated as to its authenticity. Lacking the holy grail of a holotype specimen, the field was ripe for the application of new technologies (Introduction).

It's great that geneticists have at last turned their eyes and thoughts towards hominology. The PG film is still debated as to its authenticity but only by those who are strangers to it. All who have explored it minutely do know it's authentic. Being one of those, I wrote to colleagues in 2006:

From a hominologist's viewpoint the question of Bigfoot's existence has long been solved. Bigfoot is real because the Patterson-Gimlin film is real. Period. If that is not sufficient for someone, here's another definitive statement: Bigfoot is real because some Bigfoot footprints are real. Anyone who wants to refute these statements has to prove that the film is fake and all footprints are fake. Any skeptical talk without doing this is nothing but senseless babbling. That's the ironclad logic of science. As for critics, they think the theme is grotesque and don't care for logic.

I pay so much attention to the PG film because this is the year of its jubilee and also because our priority and success in its verification show the practical worth of a good theory. Two more examples serve the same purpose. The

Porshnev theory of relict hominoids is responsible for my being the first, outside Australia, in declaring this biologically "anomalous" continent to be homin habitat (see "The Case for the Australian Hominoids" in *The Sasquatch and Other Unknown Hominoids,* pp. 101-126). I stood by this conclusion in a debate on the pages of *Cryptozoology,* Vol. 6, 1987— against the opinion of the Australian anthropologist Dr. Colin Groves, who said that "there is no substance to such a supposed beast" and "If there is a genuinely cryptozoological basis for any of the wild man (Yowie) stories, a wombat is quite certainly what it is." Journal Editor Richard Greenwell refused to publish my objection to this, saying an Australian knows better the situation there than a Russian. Dr. Colin Groves is now on the RHI Editorial Board and I wonder if he ever admitted his wrong view.

The other example of theory's influence is my study of folklore and demonology in the book in Russian, *Wood Goblin dubbed Monkey. Comparative study in demonology,* 1991. The logic and impact of the book appeared so strong that it turned Dr. Nikolai Vereschagin, hominology's worst critic and opponent, into a friend and supporter. I consider this to be my best practical achievement in this field of research.

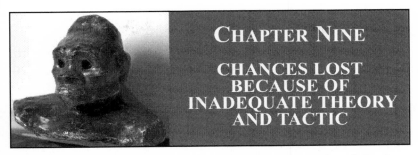

CHAPTER NINE

CHANCES LOST BECAUSE OF INADEQUATE THEORY AND TACTIC

The appearance and existence of The International Society of Cryptozoology has also been a significant factor in the making of hominology. Here is how its formation was announced:

> The International Society of Cryptozoology (ISC) was officially formed at the first (founding) meeting of the Board of Directors held in Washington, D.C. on January 8–9, 1982. The meeting was hosted by the Department of Vertebrate Zoology, National Museum of Natural History, Smithsonian Institution, whose Chairman, George R. Zug, is an ISC Board member. Dr. Bernard Heuvelmans, the acknowledged "father" of cryptozoology, was elected as President, Dr. Roy Mackal as Vice President, Mr. Richard Greenwell as Secretary/Treasurer and Editor of ISC publications.

Unexpectedly, I was invited to be a member of the ISC Board of Directors. The invitation was a great personal honor, hopefully opening some perspective for progress of hominology, but it also aroused doubts and misgivings, because the official status of cryptozoology and hominology, affecting their relationship, had not been defined and accepted yet. The founding role of the Smithsonian Institution that denied the existence of Sasquatch, was not encouraging at all.

The purpose of the Society was announced as follows:

> The ISC serves as a focal point for the investigation, analysis, publication and discussion of all matters

related to animals of unexpected form or size, or unexpected occurrence in time and space. The Society also serves as a forum for public discussion and education, and for providing reliable information to appropriate authorities.

As a matter of fact, the most prominent cryptids on the agenda of cryptozoology at the time happened to be Mokèlé-mbèmbé, a suspected dinosaur in the swamps of the Congo (a reconnaissance expedition, headed by Greenwell and Mackal, had just returned from Africa), suspected plesiosaur Nessie of Loch Ness in Scotland and the Lake Champlain Serpent, "Champ," in the US/Canada. All three figured prominently and occupied almost all space in the first volume of journal *Cryptozoology* in 1982.

As to hominology, the most conspicuous and weighty baggage of its evidence was present in the US itself, in the very backyard of the ISC and related to a single subject, the Bigfoot/Sasquatch phenomenon—thousands of Sasquatch sighting reports, hundreds of footprints, the PG film, the Sierra Bigfoot vocalizations and, last but not least, the Minnesota Iceman. All categories of this evidence were described and discussed in a dozen solid books (including *The Scientist Looks at the Sasquatch,* University Press of Idaho, 1979). In hundreds of articles, each category was crying for further study, verification and reliable information to the public and the scientific community.

The 12-member Board of Directors had only two persons engaged in the study of Sasquatch: physical anthropologist Dr. Grover Krantz and me. Grover was focused on the study of Sasquatch footprints, and the next, 1983 issue of *Cryptozoology,* carried his paper "Anatomy and Dermatoglyphics of Three Sasquatch Footprints." As for me, I was grateful to Editor Greenwell for publishing my piece "A Note on Folklore in Hominology" in the first volume of the journal, but what concerned me most was the attitude of the Society to the Patterson and Gimlin film, authenticated by Russian hominologists a decade earlier. In this respect, the first ISC publications were disappointing. This is reflected in my letter to Richard Greenwell of June 30, 1983:

> I noticed the emergence of a "Nessie lobby" in our Society. No grudge on my part unless there is also a "Bigfoot lobby," and a healthy competition between the two. Yet I see that the former has already come out publicly in defense of Nessie's rights, whereas Bigfooters haven't been given such a chance in the Society. I see that Nessie-related photo materials are given minutest consideration in the ISC publications, while the best Bigfoot film to date, with a wealth of surprising details, is being kept mum about.

Article VI of the ISC Constitution said: "The Board of Directors has the power to appoint and dissolve committees and to reassign duties thereto as may be necessary to effect policies and programs instituted by the Board." So in a letter to the Board I suggested setting up a Hominology Committee. My argument was as follows:

> Now that cryptozoologists have come together in a united front against all "infidels," they can afford to separate a little along "professional" lines so as to make more cohesion and efficiency in each branch of cryptozoology. Since zoology is subdivided into mammalogy, primatology, ornithology, herpetology, etc., it seems only logical and inevitable that cryptozoology should also have branches for various kinds of cryptids. I don't know what the nessieologists and Mokele-mbembeologists think on this score, but as a hominologist I do feel a need for a special Committee to expedite "scientific inquiry, education and communication" among people interested in the subject of relict hominoids.

Describing the work and tasks of such a Committee, I wrote the following:

The Patterson-Gimlin film is so far the best piece of photographic evidence in the whole of cryptozoology. Marcellin Agnagna's failure to film Mokele-mbembe has only emphasized the late Roger Patterson's feat. After the film's study in the USSR better quality color prints, made in the US, have even allowed us to see such details of the hominoid's anatomy as the so-called double ball on its soles, previously known only from the footprints. It is a shame and a great historical blunder that the film has not yet been studied scientifically in the US. It would be an urgent task for the Hominology Committee to take steps to correct this mistake.

The Board of Directors turned down my proposal, the argument being that it "would weaken the Society," stating: "Such a precedent could lead to moves for a Lake Monster Committee, a Giant Octopus Committee, a Mokele-Mbembe Committee, etc., since recognizing relict hominoids as worthy of such special attention could be seen as downgrading the importance of other areas of cryptozoological interest."

The argument for refusal was not convincing to me. The evidence for relict hominoids was clearly a priority and in need of urgent and special attention because it was incomparably more solid and abundant than in other areas, of different kinds; in some cases already well described or studied and in need of final verification. And yes, for me and I believe for science, a living *Homo erectus* or Neanderthal is of more importance than a living dinosaur. As John Napier warned, a Bigfoot, if real, would oblige scientists "to re-write the story of human evolution." Cryptids of different kinds do not foretell such radical revisions in science, as proved by the case of the Coelacanth "living fossil." So, yes, as proclaimed by George Orwell, "All animals are equal, but some animals are more equal than others." This kind of anthropocentrism is quite natural and inevitable for mankind bent on "knowing itself." It's just for this reason that paleoanthropology separated from paleontology and became a special discipline. A similar perspective is in store for hominology.

Clearly, the majority of the Board members had no such ideas, which was not surprising, but one negative vote was strange indeed. It came from my only fellow-hominologist, Professor Grover Krantz. He was a very interesting personality, a man of many interests, and we were good friends, engaged in lively correspondence, which touched not only on anthropology and hominology. When I told him of my unorthodox views in physics—the nature of gravitation in particular—he said that as a young man he had held the same ideas. Our discussion on problems of physics was cut short by his death. He sent me his article on the origin of the Indo-European languages, and another article on his reform of English spelling. I welcomed the latter, though was unable to judge how good and practical his reform of spelling was. When as a boy I started learning English, I heard someone say in a serious voice: "In England they write Manchester and read Liverpool." Well, there is no more irrational, inconsistent and contradictory system of communication in the world than English spelling. It works havoc, in particular, in the pronunciation of foreign personal names. Igor Bourtsev/Burtsev can tell you something about that. Quite an "advantage" of a global language. Sometimes I think that rationality is so highly valued in the Anglo-Saxon world just in protest against the irrationality of English spelling. Please excuse this involuntary aside. What else remarkable has Grover written? Oh, a most remarkable and touching story, "Only a Dog," about Clyde, his giant Irish wolfhound.

So Krantz voted against a Hominology Committee! How come? Before venturing an opinion let me quote from his book *Big Footprints: A Scientific Inquiry into the Reality of Sasquatch,* 1992:

> Clearly, the only evidence that will ever be accepted
> is a body, or a large part of one. And the skeleton is

the very best part. But even then it may well take some leg-work to impress the Scientific Establishment. I used to think that if this definitive evidence became available, all of the "great men" would promptly come to see it. Judging from their response (or lack of it) to the casts with dermal ridges, this might not be the case. Not a single person with the appropriate expertise came to my university to look at them; it was I who carried the casts across the country and around the world to seek out those experts. Two of them, an anthropologist and a primatologist, actually called the ridge detail fake from looking at photographs, without even seeing the casts. Patterson got the same reception with his film. /.../

It was Bernard Heuvelmans who told me that if I found clear skeletal evidence of the sasquatch, none of the "great men," least of all the skeptics, would come to look at it. In his opinion I would have to carry the material myself to the various authorities and professional meetings, and pester the experts with it for years before they eventually would accept it. Richard Leakey or Donald Johanson need only announce a new discovery, and many of their colleagues will gladly come to see it. The accepted scientific paradigm of today includes australopithecine fossils that are three million years old; it does not include their gigantic modern equivalent. "I'll see it when I believe it" is literally true in this case. Eugine Dubois faced that problem in the 1890s when he had to carry his *Pithecanthropus* fossils from Java to the doubting professors all over Europe. Raymond Dart got the same reception in the 1920s when he found the first *Australopithecus* in South Africa. Neither of these fossil hominids was part of the scientific paradigm of their day; but thanks to the efforts of their discoverers, they are today (pp. 254-55).

It can be seen here how different Krantz's approach and mentality were from those of Porshnev regarding the hominoid problem. No prospect of a new discipline and a revolution in science for Dr. Krantz. He mentions "The accepted scientific paradigm of today," but no denial of it by hominology, which means he hasn't learned the proper lesson from Kuhn's theory, as Boris Porshnev had. No mention of paradigm and expertise shift. It's just the latter that prompts to dispense with "leg-work to impress the Scientific Establishment." Noteworthy is also this sentence: "If we were to get a DNA sequence from future acquisition of biological material, it is by no means guaranteed that anyone outside of this specialized field would feel compelled to take notice" (p. 254). Thus, if "the only evidence that will ever be accepted is a body, or a large part of one," the task is to get it as soon as possible! A Hominology Committee would only be a hindrance and postpone the solution. Accordingly, on p. 255, in big letters, there are the words **What To Do**. Let me cite from that part of the book:

> The only way to prove that sasquatches exist is to produce a type specimen. /.../ That specimen most likely will be brought in by a hunter, hard-core or otherwise. /.../ When I speak publicly on this topic, I make the point that a sasquatch legally can be shot. There can be no punishment for shooting an animal that officially does not exist. [The sasquatch is] a mythical beast with no more legal reality than a unicorn. /.../ All prospective hunters should also be cautioned that it is illegal to shoot people who walk around the forest in fur coats. Even if they are dressed in gorilla suits at the time, which is tantamount to suicide, the charge would be some form of manslaughter. /.../
>
> If such a hunter someday succeeds in this endeavor, I have some procedural suggestions. /.../ The best procedure is to cut off the biggest piece you can carry and then go for help to retrieve the remainder. /.../ The best part to take is the [skin and] head; if this is too heavy, leave the skin behind; but at a minimum, cut out the lower jaw and

bring that back. If more than the head can be taken, get a foot; if still more, bring a hand; almost anything beyond that is about equally useful (pp. 257, 258).

The easiest way that a sasquatch can be encountered, where a hunter might take it, is while driving along a back road at night. About half of all sightings occur under these circumstances. A car could be fitted with extra lights facing partly to the sides, and a powerful gun kept at hand. One could then drive slowly along such roads from midnight until just after dawn (hunting *Homo nocturnus Linnaeus,* you know – DB). I have tried this for about seven or eight nights, on three separate occasions; many animals were seen, but no sasquatch. My usual speed was 25 miles (40 km) per hour. After some practice, I was able to stop the car, set the brakes, turn on the extra lights, pick up and load the gun, and be standing "at point" outside of the car—all in just fifteen seconds. At least half of such sightings in the past have lasted this long or longer (p. 259).

Let's note that Grover's hunting tactic has never worked. This reminds me of what people in the Caucasus told Koffmann about Almasty: "Almasty knows all about you. As soon as you leave Moscow for the Caucasus to look for him, he already knows it." Some more from Grover's book:

This kind of roadhunting should be done only by those who know exactly what they are doing. On one excursion I saw an erect biped walking by the road, with broad shoulders and no constriction at the neck. I knew immediately that it was too small and too narrow to be a sasquatch. It was a man wearing a hooded jacket. Someone else might have shot him. The procedure is not recommended for anyone except experienced hunters, a mistake of this kind cannot be retracted (p. 259).

Grover Krantz was described in the press as follows:

He is one of the most vocal proponents of killing a sasquatch in order to prove they exist. "I want to rub a few faces in the corps," he says. His advice is to shoot it, cut off an arm or a leg or anything you can carry and get the hell out (*Human Behavior*, September 1978, p. 20).

Grover did not conceal that he was strongly criticized for his views and position:

Up to then I had received considerable public criticism, including printed and verbal abuse, for suggesting that a sasquatch should be shot (p. 264).

Several classes of grade school children sent me packets of individually written letters begging me not to do this, and informing me of my errors and about my ignorance of the true situation. Many letters and phone calls came to me and to the Anthropology Department with the same message, often adding that it was me who should be shot, not a sasquatch (p. 244).

So much was actually standing behind his vote against a Hominology Committee. He admitted to me that his stance was not ethically good, to put it mildly, but he just "couldn't miss a once in a lifetime chance" for making a historic discovery. My opinion was and still remains that, on the contrary, his attitude was causing us to lose our chances in making scientific progress. What's worse, he was only "one of the most vocal proponents of killing a Sasquatch in order to prove they exist." Another one of the most vocal and influential was, of course, John Green. To wit:

Science will keep its eyes tight shut until someone produces a body, or part of one, and the more quickly it is done the better. The successful hunter should find it very profitable as well (*The Sasquatch File,* p. 71).

The difference between Krantz and Green was only in the number of Sasquatches needed to be killed for the success of our cause. Krantz mentioned one, Green accepted and predicted many:

> Should they be hunted for scientific purposes? Definitely yes. /.../ Following that, should they be captured for public display and for study? The same consideration applies to them as to other animals. /.../ The situation of the sasquatch differs considerably from that of the other great apes in at least three ways. First, there is no shortage of wild sasquatches. They cover such a tremendous area that there must be many thousands of them, and there is nothing to indicate that their numbers are declining. On the contrary, their appearance in more and more places where they were not previously known suggests that they are steadily becoming more numerous. /.../
>
> The third difference is that sasquatches are not available for study without killing them. /.../ Thorough study of all the various systems—muscles, nerves, glands, blood vessels, digestive organs and so on—requires dissection of quite a few bodies. The only way they can be obtained, in the case of a sasquatch, is by hunting (*Sasquatch: The Apes Among Us,* pp. 462-64).

It's paradoxical that such attitude toward the subjects of their search and study was displayed only by the hominological members of the International Society of Cryptozoology, but never by cryptozoologists *per se.* None of them ever as much as hinted at killing a Mokele-mbembe, a Nessie or a Champ. Also noteworthy was my failure to persuade Green and Krantz to follow a more reasonable course in this matter. They just ignored my arguments and objections and stubbornly pursued their line of action. One can read about that in the books *Bigfoot: The Problem of Proof. To Kill or To Film?,*

2000, edited and published by Pyramid Publications (Christopher L. Murphy), and *Bigfoot Research: The Russian Vision,* 2011, Hancock House.

It is also remarkable how infectious "bloodthirsty" sentiments are. From *Bigfoot Times,* December 2016, I learned the following:

> From last year on April 5, 2015, Kathy Strain addressed quite a few of us in an e-mail communication, in which she makes her point of view very clear. She writes, "To also be clear, I am pro-science... meaning I am pro-kill. I am in the Grover Krantz/John Green camp. If you have ever been to the south and witnessed the massive clear-cutting (not just conifers, but also of hard woods which we have demonstrated is part of their food resources); experienced the gigantic wildfires happening in the west, you should know that suitable habitat is dwindling at a rapid pace. Their ecosystems are changing. If we don't act, our silence is akin to criminal.

I am much satisfied that, contrary to Kathy Strain, the Bigfoot Field Research Organization (BFRO), the oldest and largest of its kind, declares the following: "It has always been the policy of the BFRO to study the species in ways that will not physically harm them." <https://www.bfro.net/>.

My view is that it is the pro-kill stand that is harmful and anti-scientific. Why should we believe Green and Krantz that a Sasquatch body or part of one would convince the skeptics and the Establishment? Krantz mentions "the new adage," i.e., "I'll see it when I believe it" (p. 72). Meteorites became a reality for science only when astronomers accepted the idea of stones falling from the sky. Sasquatches will become a reality for the scientific community only when most scientists have accepted the evidence for their existence. And this cannot be achieved by a single action and a single piece of evidence, but only by the whole weight of hominological evidence and the authority of our science. In this connection Boris Porshnev wrote:

> The public is much taken by the illusion that the "snow-man" problem can only be solved by a sensational breakthrough. A single proof will be obtained and submitted: here you are! No, the process of science is more modest and more majestic. In its course knowledge is accumulated and deepened, new information is added to old information, and its overall reliability increases. A single sensation won't work if only because any sensation can be questioned... (or just ignored and covered-up! – DB). Science operates, as a rule, not with isolated facts but with series of them (Boris Porshnev, "Comment on the Patterson-Gimlin Film," journal *Znanie-Sila,* 1968, No. 9, pp. 52, 53).

John Green and Grover Krantz have provided more evidence for the existence of Sasquatches than any other investigators. And they could have done much more, especially with the advantage and facilities of the ISC, if not for the hindrance of their pro-kill dogma. If a living dinosaur had been caught on film equal in quality to the PG film, can you imagine its being rejected and ignored for 50 years? The PG film was ignored by the International Society of Cryptozoology during the whole time of its existence, with Grover Krantz on its Board and John Green being an honorary member.

Sorry, the ISC President Bernard Heuvelmans once mentioned the Patterson and Gimlin film in this way:

> The film by the late Tim Dinsdale, revealing the surface movements of one of the Loch Ness animals, and which was carefully analyzed by experts at the Joint Air Reconnaissance Intelligence Centre (JARIC) of the Royal Air Force, is certainly more convincing than the Patterson film showing an alleged Bigfoot with some grievous anatomical inconsistencies (*Crypto-zoology,* 1988, Vol. 7, p. 15).

I was furious. John and Grover kept silent. Richard

Greenwell refused to publish my protest. Krantz says in his book: "I was just as naive as any of the younger scientists today, perhaps more so; the idea that such studies *ought not* to be published never occurred to me" (p. 241). Professor Krantz seems to have remained somewhat naive regarding some aspects of hominology to the end of his life. His corresponding theory and hypothesis are unsophisticated and even simplistic. What was his argument for the choice of *Gigantopithecus blacki* as Sasquatch ancestor? Essentially the gigantic body size. But it's like deciding that the ancestors of his gigantic wolfhound were gigantic wolves. *Gigantopithecus* was an ape, so Sasquatch is also an ape. A legitimate object of killing. "It is not human, nor even semi-human, and its legal status would be that of an animal if and when a specimen is taken" (p. 173). There's also this marvelous admission by Krantz: "I call the Sasquatch an ape, though it most probably is a hominid" (*Bigfoot Research,* p. 276). And this: "This one may well be the most important of all such unproven animals because it is probably our closest living relative" (p. 272). Consistent thinking, isn't it? And finally this:

> It might be argued that we don't really know enough about sasquatch behavior to be absolutely certain about this judgment as to its animal status. But if we are in error, isn't it imperative that we find out as soon as possible? (p. 12)

Find out how? By killing one? Bravo! As to Green, he was not far behind Krantz on this kind of theory and tactic; he may have been even ahead. In a published article "What is the Sasquatch?" he answered, "the Sasquatch is an animal—an upright ape—and nothing more." His big important volume is titled *Sasquatch: The Apes Among Us,* 1978. His lead was taken by others: John Bindernagel, *North America's Great Ape: The Sasquatch,* 1998; Loren Coleman, *Bigfoot: The True*

Story of Apes in America, 2003. I call it the Ape Syndrome.

Back to the PG film. In 2000, I contacted by email Professor Philip Lieberman at Brown University, specialist on the evolution of the biological bases of human language, and among other things asked for his opinion of the Patterson and Gimlin film. He answered, "The supposed Bigfoot film appeared to primate specialists to be that of a human walking, wearing a crudely modified ape costume." That was 33 years after the film was taken and 28 years after it was authenticated in Russia. The situation hasn't changed in the US since then.

What about Russia? In 2005, at the ceremonial opening of the Darwin Museum's new building, I had a chat with a well-known journalist, Vasily Peskov, author of a newspaper column "A Window on Nature," in which he told of the great value and achievements of photography in showing the wonders of wildlife (see *America's Bigfoot,* p. 7). He wrote: "So never fail to give credit to the camera. Doubtless, there is still many a mystery in Nature to be seen and confirmed by the camera." With reference to these words, I mentioned the PG film, which opens one's eyes on a great mystery of nature. Peskov responded that the film was a hoax. "How do you know that?" I asked. "Well, it's known that Americans themselves have exposed it," he said, and declined any further discussion.

The situation remains the same in Russia as well. What a shame. Who or what is to blame? No matter who or what, it is clear that a lot of chance and time has been lost by American hominologists because of inadequate theory and tactic.

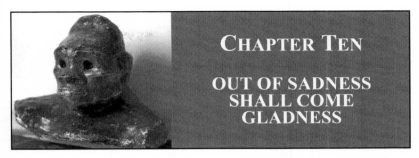

CHAPTER TEN

OUT OF SADNESS SHALL COME GLADNESS

Albert Ostman asked his Indian guide what animal he called a Sasquatch and the guide said: "They have hair all over their bodies, but they are not animals. They are people. Big people living in the mountains." A few days later Ostman learned that the guide was right.

We all know and respect a man who saw a Sasquatch very clearly at close quarters—face, body, movement and gait— and we all saw that Sasquatch on film. The man is Robert Gimlin who said: "I'm not going to call it an animal because I don't believe it is" (Jeff Meldrum, *Sasquatch,* p. 141).

We have testimony enough now and even some data (the Sierra vocalization recordings) that Sasquatches have language, the unique characteristic of humans. By this criterion they are definitely not animals but people. This means that not only John Green and Grover Krantz were in error regarding the nature of Sasquatches, but the founding father of hominology itself, Professor Boris Porshnev. That was quite a blow to my self-confidence for I had accepted for decades Porshnev's theory regarding the nature of relict hominoids and spread it worldwide. True, his belief that all hominids different from *Homo sapiens* were animals caused me what is termed nowadays as *cognitive dissonance*. Therefore I used to regard and describe them as *super animals*. Still, radical change began to take place in my thinking only in 2002, after acquaintance with Mary Green and Janice Carter and reading their book *50 Years with Bigfoot: Tennessee Chronicles of Co-Existence.* As a result I wrote:

85

Bigfoots of North America were believed by all investigators, myself included, to be less humanlike than their counterparts in Asia and Europe. If they turn out to be a kind of human, as it follows from what we learn from Albert Ostman and Janice, then Bigfoot counterparts in Eurasia are also human. The universal historical term "wild man" may happen to be literally correct. If recognition of the existence today of non-*Homo sapiens* hominids means a revolution in scientific knowledge, then recognition of those hominids as humans will mean a revolution in our knowledge of the subject. A new chapter, nay, a new stage of hominology will have to begin, demanding a radical change of our thinking and our approach to the matter (*Bigfoot Research: The Russian Vision,* p. 401).

Science, as said in the beginning, is humanity's main Hunter for Truth. Truth means being in accord with fact and reality. Porshnev's theory was practically good, but as it is seen now, not good enough. Science isn't perfect. Scientists make mistakes. Science is self-correcting. Science may take wrong turns from time to time, but it eventually finds its way back on the right road. Doing the correction, let me remind you of Sanderson's words:

These Soviet activities shed an entirely new light on the whole business ... /.../ This was that the whole problem is an anthropological rather than a zoological matter. In other words, all the Sino-Soviet evidence pointed to ABSMs being primitive Hominids (i.e., Men) rather than Pongids (i.e., Apes) or other nonhuman creatures ...

And let us recall Porshnev's words: "I would never have been involved in the problem of snowman had I thought it was an ape." Thus the difference between the error made by Green, Krantz, Bindernagel and others in North America and that of Boris Porshnev and his followers in Russia, including myself, was in the fact that Russian hominology has never suffered from the Ape syndrome, so prevalent and detrimental in North America.

Hominology came into being in a no-man's land of science between zoology and anthropology. It has been shifting ever since from the zoological side of the area to the anthropological side. Now that it has reached that side, it is necessary to look again at hominology's main problem in the opinion and words of Grover Krantz:

> Science will not accept the sasquatch as a living species without a type specimen … (p. 246). The only way to prove that sasquatches exist is to produce a type specimen. Virtually every established scientist will repeat this same demand (p. 257)

By the categorical nature of these statements Krantz convinced himself and many others of their absolute veracity. So let us take a closer look at them. They casually touch on several important concepts, such as existence, proof, science, and a technical concept and term "type specimen." The question is: Will science in general or only some scientific discipline "not accept the Sasquatch without a type specimen?" A similar question refers to "virtually every established scientist."

From the experience of René Dahinden in Moscow, from the case of Dr. Dmitri Donskoy and many other cases, we do know about different attitudes of scientists of different scientific disciplines to the relict hominoid problem, the Bigfoot/Sasquatch phenomenon in particular. It is also worth quoting Grover on the matter of type specimen:

> New animal species become generally accepted when a physical specimen is collected and properly described. Most commonly, living species are initially demonstrated by the skin and skull of one individual. Extinct species are based on fossilized bone or shell, often on rather small fragments. Some are proposed even on the

basis of such indirect evidence as burrows or tracks. In all cases, something tangible is required, which is deposited in a collection somewhere, and which can be examined later by persons other than the discoverer. /.../ There are no clear rules about the kind and amount of evidence that must be provided in each case. /.../ ... the first platypus skin brought to Europe was suspected of being faked (Grover S. Krantz, "Anatomy and Dermatoglyphics of Three Sasquatch Footprints," *Cryptozoology*, 2, 1983, p. 53).

As a matter of fact, faking is applicable not only to skin, but to skeletal evidence as well. A somewhat "desacralizing" example of such kind of evidence is, of course, the Piltdown skull. The following is from Wikipedia:

The Piltdown Man was a paleoanthropological hoax in which bone fragments were presented as the fossilized remains of a previously unknown early human. /.../ The Piltdown hoax is prominent for two reasons: the attention it generated around the subject of human evolution, and the length of time (45 years) that elapsed from its alleged initial discovery to its definitive exposure as a composite forgery.

So much for Professor Krantz's advice "take the head ... at a minimum, cut out the lower jaw ..." as an unequivocal solution of the problem.

If that is not enough to stop exaggerating the importance of the type specimen in science, of its being the criterion of what is real and what is not, let us note that absolutely all animals actually existed (or exist) before their existence was or is registered by means of a type specimen. They existed or exist *de facto*, while the type specimen procedure only puts a *de jure* "stamp" on this fact. This "stamp" formality is required in zoology, paleontology, paleoanthropology and botany.

Thank goodness, we can forget about the type specimen procedure, at least for the time being, because hominology is no part of zoology, paleontology, paleoanthropology and botany. As we perceive now, it is part of anthropology! It can't be anything else if Sasquatches are people, not animals. In anthropology you don't necessarily need to present a type specimen as proof of reality. Linnaeus did not designate a type specimen for *Homo sapiens*, nor *Homo troglodytes*.

Unlike every other human species, *Homo sapiens* does not have a true type specimen. In other words, there is not a particular *Homo sapiens* individual that researchers recognize as being the specimen that gave *Homo sapiens* its name. Reference: Homo sapiens – The Smithsonian Institution's Human Origins Program: <http://humanorigins.si.edu/evidence/human-fossils/species/homo-sapiens>.

Homo sapiens is taken for real without any type specimen. Accordingly, nobody ever demands skeletal evidence for the existence of, say, Plato and Aristotle or Caesar and Cleopatra.

Now, anthropology is subdivided into several fields of science and the first and main division is in physical (biological) and cultural anthropology. Grover Krantz, being a physical anthropologist, regarded and studied Sasquatches from the viewpoint of physical anthropology. This was correct and proper first of all in regard to one category of evidence, namely footprints, which represent the anatomy of Sasquatch feet. As to some other kinds of evidence, their data belong to cultural, not physical anthropology. So hominology is standing with one leg in physical anthropology and the other in cultural anthropology. Let me note and stress immediately that this does not detract in the least from the reality of homins, **as is clear from what is said above.** Cultural anthropology has its own criteria of reality and they are as necessary and scientific as those of physical anthropology.

Analogy and comparison are legitimate and usual methods of cognition in science. Linnaeus and Darwin compared man with monkeys and came to important conclusions. Dr. Bindernagel and others compared Sasquatches with gorillas and concluded that Sasquatches are bipedal apes. Finding this conclusion incorrect, I compare homins with bipedal primates that are called human beings or people. And the following is what I found in Wikipedia as some analogy of the Bigfoot/Sasquatch phenomenon:

> Uncontacted peoples, also referred to as isolated peoples or lost tribes, are communities who live, or have lived, either by choice (peoples living in voluntary isolation) or by circumstance, without significant contact with global civilization. Few peoples have remained totally uncontacted by global civilization. /.../ [Nevertheless] in 2013 it was estimated that there were more than 100 uncontacted tribes around the world, mostly in the densely forested areas of South America, Central Africa, and New Guinea. **Knowledge of the existence of these groups comes mostly from infrequent and sometimes violent encounters with neighboring tribes, and from aerial footage.**
> (My emphasis – DB)

Note that scientists, namely anthropologists, do know of the *existence* of these tribes without any type specimens; just by means of witness accounts and film footage, which is also the practice of hominologists. One significant difference though between homins and "uncontacted peoples" is the fact that homins exist not only in the wilderness of densely forested areas of the world, but also within populated territories of many countries, still remaining unrecognized and unregistered by established modern science.

This is due to their ability to avoid detection by *Homo sapiens* when it is undesirable, which surpasses the abilities of any animals. Bigfoot "is never seen unless he wants to be seen or doesn't care. He is, as Archie Buckley points out, a 'master of concealment.' He can approach a camp, even in daylight, and never be detected" (George Haas).

What's more, they successfully avoid detection by photo-traps, which is on the verge of miraculous. At the 1978 Vancouver Sasquatch Conference, Dr. Jim Butler presented his report "The Theoretical Importance of Higher Sensory Perceptions in the Sasquatch Phenomenon." This is a topic of great importance for research to come when hominology is properly established and funded. But even today we can recognize its legitimacy by comparing with still scientifically unexplained abilities of such human bipeds as yogis. References:

Harvard goes to the Himalayas – Monks with "Superhuman" abilities show scientists what we can all do.

<http://www.collective-evolution.com/2016/03/01/harvard-goes-to-the-himalayas-monks-with-superhuman-abilities-show-scientists-what-we-can-all-do/>

Attaining the Siddhis: A Guide to the 25 Yogic Super-powers.

<http://www.consciouslifestylemag.com/siddhis-attain-yoga-powers/>

In short, the sadness of hominology's condition within cryptozoology is to be changed for gladness and successful research within anthropology.

CHAPTER ELEVEN

MOVING IN THE RIGHT DIRECTION

The making of hominology has turned out to be a long, contradictory and precarious process. Firstly, this is because hominology brings a paradigm change, acceptance of which is a "painful" and protracted procedure for science. Secondly, there has not been enough cohesion and cooperation among hominologists themselves and this holds back real progress. The title of my last book is *Russian Hominology,* which is not, strictly speaking, correct, if hominology claims to be a science. Science is not divided by ethnic or geographic criteria. This means that the discipline is not yet fully formed and mature. The fact is that research in Russia and North America has been based on and guided by different theoretical concepts which I described earlier. Russian Hominology actually implies hominological research done by Russian investigators. What we need though is that research in this field, both in Russia and America, and in any country, be done within a single scientific discipline based on universally approved principles.

Fortunately, in North America there have already been steps and movement in this direction. To begin with, back in 2001 and 2002, Craig Heinselman, of the USA, published two collections of papers, titled *Hominology Special Number I* and *Hominology Special Number II* presenting, among others, such papers as "Early Man as a Model for Bigfoot" by Ray Crowe; "What is Living in the Woods, and Why it isn't *Gigantopithecus*" and "Predictability of Homin Behavior" by Will Duncan. The year 2015 saw publication of *Bigfoot in*

Evolutionary Perspective: The Hidden Life of A North American Hominin, by T.A. Wilson. The book ends with this statement:

> For centuries man has pondered whether he is alone in the universe. It seems a revision is in order. It is time to ask if he is even alone on the planet. This is where the needle points (p. 336).

Chris Murphy has authored and David Hancock has published two encyclopedic tomes: *Meet the Sasquatch,* 2004, and *Know the Sasquatch/Bigfoot,* 2010, presenting side-by-side the evidence and findings by the multidisciplinary investigators of the United States, Canada and Russia. In addition, Murphy and Hancock have done their best to acquaint North American readers, through book publications, with the work of Russian hominologists. Still more, unique and most valuable work is being done by Chris Murphy by curating major Sasquatch exhibits in public museums in Canada and the United States, which have won popularity in these countries.

Chris tells me apologetically that his exhibits, presenting on view all sorts of objects of hominological evidence, are meant to be *cultural,* not *scientific* events, which view I find to be a misconception. As said earlier, hominology is part of anthropology, and its database includes items of both physical and cultural nature. So, definitely, Sasquatch exhibits are important scientific (and cultural) events.

In recent years Chris has partnered with the website Sasquatch Canada, created and maintained by Candy Michlosky <http://www.sasquatchcanada.com/> and has made this site a major platform for the presentation of Sasquatch and other hominoid research. The site includes a Virtual Museum—essentially everything related to hominology gathered over 25 years is presented in "galleries." Furthermore, Chris' *Catalog of Exhibit Items* (artifacts and other material provided for physical exhibits) is featured on the site.

Of special significance was the appearance in 2012 at

Idaho State University of the online journal Relict Hominoid Inquiry. Of course, it's still far from what hominology needs and deserves, but so far it's the best academic representative of the discipline in North America.

Last, but not least, is Dr. Jeff Meldrum's presentation "Sasquatch & Other Wildmen: The Search for Relict Hominoids," delivered on June 20, 2016, at the Meeting of the Society for Scientific Exploration, held in Boulder, Colorado. It was on the occasion of the bestowing to Meldrum of the 2016 Tim Dinsdale Award for his:

> ... significant contribution to our understanding of the possible presence of an as-yet unrecognized primate in our midst. In the course of more than two decades, while recognizing the risk to his professional reputation, he has created a corpus of credible work by conscientiously applying his knowledge of primate evolutionary anatomy and behavior to this most difficult and controversial subject.

It is most significant that the words *Gigantopithecus* and "cryptozoology" are not present in Meldrum's text at all, nor any trace of the Ape syndrome. The main term used by the author for the primates of our interest is "hominin," which earlier meant "hominid." The Single Species Hypothesis is criticized and the Bushy Hominin Tree, with lingering populations of relict species is offered instead. In short, conceptually and terminologically, the problem of Sasquatch & Other Wildmen, as presented by Dr. Jeff Meldrum, quite tallies with what could be called the mainstream concepts of hominology.

Also notable is the fact that the author relies both on physical and cultural anthropology in his paper. As a specialist in physical anthropology, he tells about his extensive work on the anatomy of the Sasquatch foot, as revealed in footprints; as to evidence of cultural anthropology, namely ethnography, Meldrum refers to the work of Dr. Gregory Forth, who uses such words as "hominological reality."

In conclusion of this section, here are a few words regarding a comparatively recent novelty in fieldwork, showing the worldwide similarity and consistency of hominological subject matter. I recall being surprised and impressed by the fact that Sasquatches are as much prone to braid horses' manes as our Almasty. Even more widely spread, or at least widely registered, is the homin trait of tree manipulation. This was noticed long ago, but only recently put to good use, allowing to detect homin presence where footprints are rarely seen. Wooden constructions, some requiring inhuman strength and others great hand dexterity, have been registered in North America, Russia and Australia. Some quite close to big cities, such as Moscow, for instance, which was beyond imagination decades ago.

Another international homin characteristic is their use of different objects, such as stones, sticks, etc., to lay out on earth certain patterns which seem to be symbolic and meaningful. Igor Burtsev is devoting much attention to this intriguing enigma, causing investigators a lot of head-scratching.

CHAPTER TWELVE
GETTING TO SQUARE TWO

With all that, according to Chris Murphy, "We are still ESSEN-TIALLY at square one." This can hardly be denied if we consider "List of topics characterized as pseudoscience –Wikipedia," which mentions such names as Bigfoot, yeren, yeti, within the topic characterized by the derogatory term pseudoscience.
<https://en.wikipedia.org/wiki/List_of_topics_characterized_as_pseudoscience>

> Cryptozoology – search for creatures that are considered not to exist by most biologists. Well known examples of creatures of interest to cryptozoologists include Bigfoot, Yeren, Yeti, and the Loch Ness Monster.
> <https://en.wikipedia.org/wiki/Cryptozoology>

We see that the term and notion of hominology have not yet even reached the horizon of Wikipedia. How about the late International Society of Cryptozoology? It was officially and joyfully formed, and existed for over a decade as a widely respectable scientific body, with a lot of PhD scientists as its members and a sufficient number of sponsors and benefactors ensuring its funding. What was the Society's theoretical basis and justification? Actually, only two books: *On the Track of Unknown Animals* by Bernard Heuvelmans and *Abominable Snowmen: Legend Come to Life* by Ivan Sanderson. Where are the now telling results of the ISC cryptozoological activity *per se?*

Compare with the current tremendous body of theoretical and practical work by hominologists, with tons of evidence of various kinds, with tens of books and hundreds of papers written as a result, and a Bigfoot documentary film in the bargain; all of that with irregular and spotty funding or no funding at all!

Let me quote Dr. Grover Krantz again:

What is said here about scientific ignorance regarding the Patterson film is equally true for the footprint evidence and the testimony of eyewitnesses. None of this is normally published in the scientific journals... I don't know of a single scientist who has firmly denied the existence of the Sasquatch on the basis of a reasonable study of the evidence (*Big Footprints,* p. 123).

That was published in 1992. What is happening now? Says Dr. Jeff Meldrum:

In response to persistent indications of mysterious hominoids, we have witnessed a recent rash of skeptical books published on the subject of Bigfoot (e.g., Long 2004, Daegling 2004, Buhs 2009, McLeod 2009, Nickell 2011, Loxton & Prothero 2013). Some of these titles, penned by fellow academicians, have been inexplicably published by prestigious university presses, e.g., University of Chicago Press and Columbia University Press (*Sasquatch & Other Wildmen*).

One thing is clear and explicable: a radical difference between the subjects of cryptozoology and hominology in the minds of academicians. A plesiosaur in Loch Ness or live mammoth in Siberia would be a sensation in zoology; but Bigfoot means a paradigm shift and a revolution in science. Searches for lake and sea monsters are fun and tolerable, but the truth of the existence of Bigfoot must be banned by all means in order to defend, as aptly said earlier, the Establishment's right "to be wrong."

Turning things upside down, John Napier wrote that scientists have taken a bad beating over the subject of monsters. They are in fact the whipping boys of the monster establishment. These enthusiasts hold that scientists are suppressing information (*Bigfoot,* p.14).

Who suppresses certain information in science and why it

is suppressed is revealed by Thomas Kuhn, not "the monster establishment." Napier makes fun of the idea that scientists "have formed a conspiracy of silence." In fact, it's more than a conspiracy of silence. It's also a conspiracy of ignorance, misinformation and discrimination. So let's see who's who in this process of evil-doing and separate innocent people from culprits. Was Dr. Lieberman, who believed that the subject of the PG film was a human "wearing a crudely modified ape costume," a wrong-doer? Of course not. Was journalist Vasily Peskov, who said that Americans themselves "exposed" the film, a culprit? Definitely not. They both, like millions (perhaps billions) of people in the word were duped by the mass media which has constantly been spreading the misinformation and bare lies of real culprits in this matter—the academicians who have penned and published their libelous and pseudoscientific critique, such as pointed out above by Dr. Jeff Meldrum.

We may ask what is the government position and policy in the case of relict hominoids. Boris Porshnev was convinced that the Soviet Government, its KGB (secret service) especially, well knew about the existence of hairy wildmen, and the frontier guards had instructions how to deal with them when caught crossing the state border. All such information was classified.

What about the United States? My brief communication with the then President Bill Clinton is rather interesting in this respect. In February 1999, I sent him my books, *In the Footsteps of the Russian Snowman* and *America's Bigfoot: Fact, Not Fiction. US Evidence Verified in Russia,* and asked to do "something to defend the good name of the late Roger Patterson and the honor of Robert Gimlin. They are both real heroes of science." In August of that year, I received the following reply:

Thank you for your kind gift and for sharing your thoughts and concerns. It's important for me to know your views. I'm glad you took the time to write. Bill Clinton (*Bigfoot Research*, 2011, p. 421).

As I interpret his response, even the President of the US is powerless in this matter. This is proved by his silence regarding Patterson and Gimlin. I'm sure he would have been glad to defend them, but their names were unmentionable for him. Just as the reality of the Bigfoot/Sasquatch phenomenon. He knew of its existence, but the information is top secret and probably not fully available even to the President. Even if available, it's one-sided, and that is why my views were important for him. The very fact that I received a reply, and it was worded in that way, makes me sure that my interpretation of it is correct.

This gives some idea how powerful the anti-Bigfoot lobby is in the US. No biological evidence, including genetic, will be taken into consideration. It will just be ignored, denied, hushed up, confiscated, whatever; but definitely not recognized.

So no way out? As usual, there is only one way out under such circumstances: Build up the discipline which can alone expose and defeat the enemies of truth in science. Our word is based on knowledge, theirs on ignorance. But that is not yet apparent to journalists, the reporters and news makers of the mass media. They all well know the word "Watergate" and must learn now the word "Bigfootgate." The former means a great cover-up and scandal in party politics; the latter the greatest cover-up and scandal in modern science history. The task is to make this apparent to the journalist community; to win over honest reporters so that they can deal with Bigfootgate the way they dealt with Watergate.

SPECIAL NOTATION: The material in the following Chapters was provided by my associate author, the sources are indicated. It may not reflect my opinion in some cases. DB

CHAPTER THIRTEEN

THE FOOTPRINT EVIDENCE

EVALUATION OF ALLEGED SASQUATCH FOOTPRINTS AND
INFERRED FUNCTIONAL MORPHOLOGY (1999)

(*American Journal of Physical Anthropology* Suppl. 27:161 – Abstract)

D. Jeffrey Meldrum, PhD

Introduction

Throughout the twentieth century, thousands of eyewitness reports of giant bipedal apes, commonly referred to as "Bigfoot" or "Sasquatch," have emanated from the montane forests of the western United States and Canada. Hundreds of large humanoid footprints have been discovered and many have been photographed or preserved as plaster casts. As incredible as these reports may seem, the simple fact of the matter remains: the footprints exist and warrant evaluation. A sample of over 100 footprint casts and over 50 photographs of footprints and casts were assembled and examined, as well as several examples of fresh footprints.

Tracks in the Blue Mountains: The author examined fresh footprints firsthand in 1996, near the Umatilla National Forest, outside Walla Walla, Washington. The isolated trackway comprised in excess of 40 discernible footprints on a muddy farm road, across a plowed

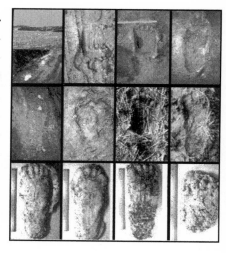

field, and along an irrigation ditch. The footprints measured approximately 35 cm (13.75 inches) long and 13 cm (5.25 inches) wide. Step length ranged from 1.0–1.3 m (39–50.7 inches). Limited examples of faint dermatoglyphics were apparent, but deteriorated rapidly under the wet weather conditions. Individual footprints exhibited variations in toe position that are consistent with inferred walking speed and accommodation of irregularities in the substrate. A flat foot was indicated, with an elongated heel segment. Seven individual footprints were preserved as casts.

Evidence of a Midtarsel Break: Perhaps the most significant observation relating to the trackway was the evidence of a pronounced flexibility in the midtarsal joint. Several examples of midfoot pressure ridges indicated a greater range of flexion at the transverse tarsal joint than permitted in the normal human tarsus. This is especially manifest in the footprint shown below, in which a heel impression is absent. Evidently, the hindfoot was elevated at the time of contact by the midfoot. Due to muddy conditions, the foot slipped backward, as indicated by the toe slide-ins, and a ridge of mud was pushed up behind the midtarsal region.

 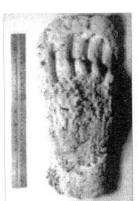 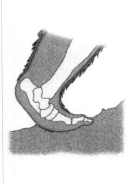

Patterson and Gimlin Film Subject: On Friday, October 20, 1967, Roger Patterson and Bob Gimlin claimed to have captured on film a female Bigfoot retreating across a gravel sandbar on Bluff Creek in northern California. The film provides a view of the plantar surface of the subject's foot, as well as several unobstructed views of step cycles. In addition to a prominent elongated heel, a mid-tarsal break is apparent during midstance, and considerable flexion of the midtarsus can be seen during the swing phase. The subject

(© L. Laverty)

left a long series of deeply impressed footprints. Patterson cast single examples of a right and a left footprint.

Three days later (Monday, October 23, 1967) the site was visited by Robert Lyle Laverty, a timber management assistant, and his survey crew. Laverty took several photographs, including one of a footprint exhibiting a pronounced pressure ridge in the midtarsal region. This same footprint, along with nine others in a series, was cast six days later (Sunday, October 29, 1967) by Bob Titmus, a Canadian taxidermist.

A model of inferred skeletal anatomy is proposed here to account for the distinctive midtarsal pressure ridge and "half-tracks" in which the heel impression is absent. In this model, the Sasquatch foot lacks a fixed longitudinal arch, but instead exhibits a high degree of midfoot flexibility at the transverse joint. Following the midtarsal break, a plastic substrate may be pushed up in a pressure ridge as propulsive force is exerted through the midfoot. An increased power arm in the foot lever system is achieved by heel elongation as opposed to arch fixation.

Additional Examples of "Half-Tracks": A number of additional examples of footprints have been identified that exhibit a midtarsal break, either as a pronounced midtarsal pressure ridge or as a "half-track" produced by a foot flexed at the transverse tarsal joint. Each of these examples conforms to the predicted relative position of the transverse tarsal joint and elongated heel. The first example is documented by a set of photographs taken by Don Abbott, an anthropologist from the British Columbia Museum (now Royal Museum), in August 1967. These footprints were part of an extended trackway, comprising over a thousand footprints, along a Blue Creek Mountain road in northern California.

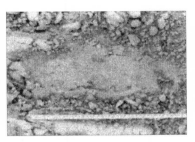

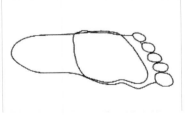

Deputy Sheriff Dennis Heryford was one of several officers investigating footprints found by loggers on the Satsop River, Grays Harbor County, Washington, in April 1982 (area is known as Abbott Hill). The subject strode from the forest across a logging landing, then, doubling its stride, left a series of half-tracks on its return to the treeline. Note the indications of the fifth metatarsal and calcaneocuboid joint on the lateral margin of the cast. The proximal margin of the half-track approximates the position of the calcaneocuboid joint.

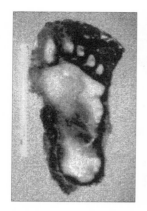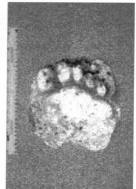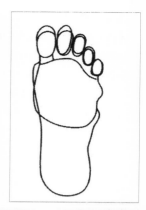

Examples of Foot Pathology: The track of an individual with a presumed crippled foot was discovered in Bossburg, Washington in 1969. The malformed right foot has been previously misidentified as a case of *talipes equinoverus* (clubfoot). However, it is consistent with the general condition of *pes cavus,* specifically metatarsus adductus or possibly skew foot. Its unilateral manifestation makes it more likely that the individual was suffering from a lesion on the spinal cord rather than a congenital deformity. Regardless of the epidemiology, the pathology highlights the evident distinctions of skeletal anatomy. The prominent bunnionettes of the lateral margin of the foot merit the positions of the cal-

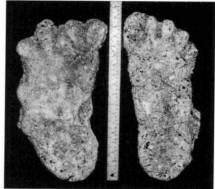

(© C. Murphy)

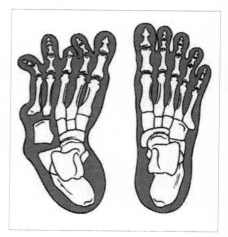

caneocuboid and cuboideo-metatarsal joints, which are positioned more distal than in a human foot. This accords with the inferred position of the transverse tarsal joint and confirms the elongation of the heel segment. Furthermore, deformities and malalignments of the digits permit inferences about the positions of interphalangeal joints and relative toe lengths, as depicted in the reconstructed skeletal anatomy shown.

Relative Toe Length and Mobility: Variations in toe position are evident between footprints within a single trackway, as well as between individual subjects. In some instances, the toes are sharply curled, leaving an undisturbed ridge of soil behind toe tips resembling "peas-in-a-pod." In other instances the toes are fully extended. In either case, the toes appear relatively longer than in humans. Among the casts made by the author in 1996 is one in which the toes were splayed, pressing the fifth digits into the sidewalls of the deep imprint, leaving an impression on the profile of these marginal toes. This is the first such case that I am aware of. Expressed as a percent of the combined hindfoot/midfoot, the Sasquatch toes are intermediate in length between those of humans and the reconstructed length of australopithecine toes. Furthermore, the digits frequently display a considerable range of abduction.

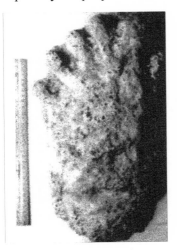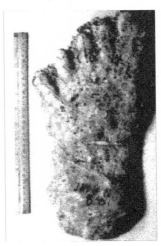

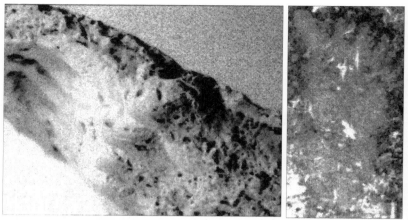

Left image shows the profile of the fifth toe on a half-track cast taken by the author outside of Walla Walla, Washington in 1996.

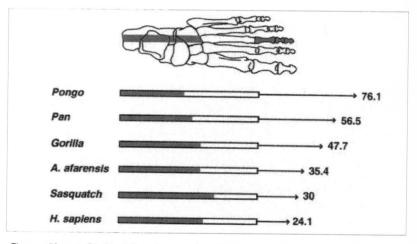

Pongo	76.1
Pan	56.5
Gorilla	47.7
A. afarensis	35.4
Sasquatch	30
H. sapiens	24.1

Compliant Gait: The dynamic signature of the footprints concurs with numerous eyewitness accounts noting the smoothness of the gait exhibited by the Sasquatch. For example, one witness stated, "it seemed to glide or float as it moved." Absent is the vertical oscillation of the typical stiff-legged human gait. The compliant gait not only reduces peak ground reaction forces, but also avoids concentration of weight over the heel and ball, as well as increasing the period of double support.

Human walking is characterized by an extended stiff-legged striding gait with distinct heel-strike and toe-off phases. Bending stresses in the digits are held low by selection for relatively short toes that participate in propulsion at the sacrifice of prehension. Efficiency and economy of muscle action during distance walking and running are maximized by reduced mobility in the tarsal joints, a fixed longitudinal arch, elastic storage in the well-developed calcaneal tendon, plantar aponeurosi, and deep plantar ligaments of the foot.

In contrast, the Sasquatch appear to have adapted to bipedal locomotion by employing a compliant gait on a flat flexible foot. A degree of prehensile capability has been retained in the digits by maintaining the uncoupling of the propulsive function of the hindfoot from the forefoot via the midtarsal break. Digits are spared the peak forces of toe-off due to compliant gait with its extended period of double support. This would be an efficient strategy for negotiating the steep, broken terrain of the dense montane forests of the Pacific and intermountain west, especially for a bipedal hominoid of considerable body mass. The dynamic signatures of this adaptive pattern of gait are generally evident in the footprints examined in this study.

Presented here are some examples of plaster casts made of what are believed to be sasquatch footprints, as collected and presented by Christopher L. Murphy. Not all have undergone rigorous scientific analysis.

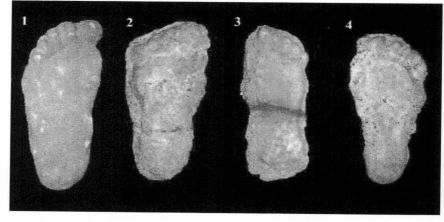

1. Bluff Creek, California, Jerry Crew, 1958 (2nd-generation cast, 17.5 inches [44.5 cm] long). This is a copy of the famous cast Jerry Crew took to a newspaper, and the resulting article gave birth to the word "bigfoot" as the name of the creature in the United States.
2. Blue Creek Mountain road, Bluff Creek area, California, John Green, 1967 (original cast, 15 inches [38.1 cm] long).
3. Blue Creek Mountain road, Bluff Creek area, California, John Green, 1967 (original cast, 13 inches [33 cm] long).
4. Believed to be from Bluff Creek, California. The person who made the cast is not known. It was probably made in the late 1960s (appears to be an original cast, 14.5 inches [36.8 cm] long).

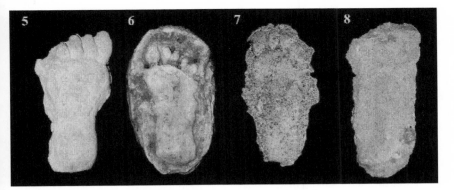

5. Strathcona Provincial Park, Vancouver Island, British Columbia, Dr. John Bindernagel, 1988 (1st-generation cast, 15 inches [38.1 cm] long). The horizontal lines on this cast were caused by a hiker who stepped in the footprint.

6. Abbott Hill, South Olympic Peninsula, Washington, A.D. Heryford, 1982 (2nd-generation cast, 15 inches [38.1 cm] long). Certainly one of the best casts ever obtained; the copy seen here was professionally produced from a mold by Richard Noll, Edmonds, Washington.

7. Shawnee State Park, Ohio, Joedy Cook, June 18, 2003 (original cast, 15 inches [38.1 cm] long). A man and his wife found the prints and called a bigfoot hotline. Cook responded and found nine footprints.

8. Chilliwack River, British Columbia, Thomas Steenburg, 1986 (2nd-generation cast, 18.5 inches [47 cm] long. Steenburg was informed of a sighting in the area three days after the occurrence and went to investigate. He independently found 110 footprints all approximately 18 inches [45.7 cm] long.

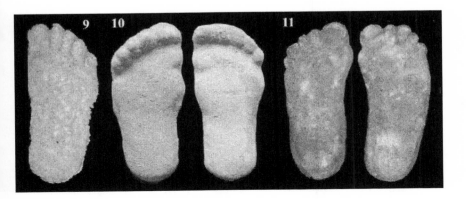

9. Laird Meadow Road, Bluff Creek area, California, Roger Patterson, 1964 (3rd-generation cast, 16 inches [40.6 cm] long). Prints were found by Pat Graves, October 21, 1963, who told Roger Patterson of the location. The sasquatch that made the prints is believed to be the same as the one that made the prints found by Jerry Crew (see No. 1).

10. Bluff Creek, California, Bob Titmus, 1958 (2nd-generation casts, 16 inches [40.6 cm] long). Both casts are from the same trackway.

11. Skeena River Slough, Terrace, British Columbia, Bob Titmus, 1976 (2nd-generation casts, 16 inches [40.6 cm] long). Both casts are from the same trackway. Children found and reported the footprints; Titmus investigated and made the casts.

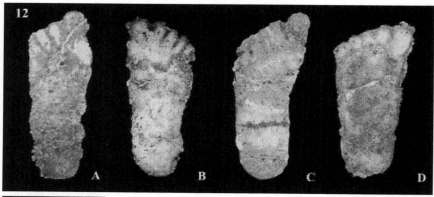

12. A–E Hyampom, California, Bob Titmus, 1963. Hyampom is a tiny village about 60 miles (96.5 km) south of Bluff Creek. All prints from which these casts were made were found on the same occasion, but only the first three prints (casts A–C, which were from the same trackway) were found in the same place. The other two casts (D and E) were from prints found in an additional two separate locations.

A. Original cast, 16 inches (40.6 cm) long
B. Original cast, 17 inches (43.2 cm) long
C. Original cast, 16 inches (40.6 cm) long
D. Original cast, 16 inches (40.6 cm) long
E. Original cast, 15 inches (38.1 cm) long

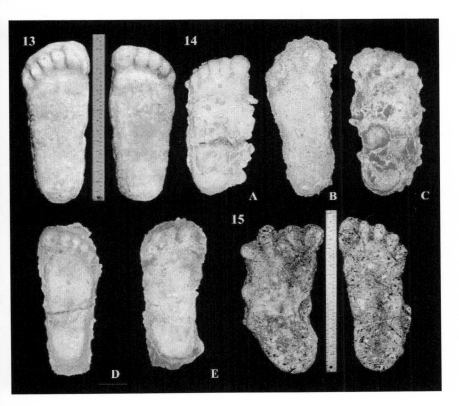

13. Patterson/Gimlin film site, Bluff Creek, California, Roger Patterson, October 20, 1967, (1st-generation casts: left cast, 15 inches [38.1 cm] long;
right cast 14.6 inches [37.1 cm] long).

14A–E. Patterson and Gimlin film site, Bluff Creek, California, Bob Titmus; from prints still in place nine days after the filming. All casts are originals. They vary in size (14–15 inches [35.6 – 38.1 cm] long.

15. Bossburg, Washington, "cripple-foot" casts (note deformed right foot – left image) René Dahinden, 1969, original casts: left cast, 16.75 inches [42.6 cm] long; right cast, 17.25 inches [43.8 cm] long). Over 1,000 footprints were found. They were discovered on two different occasions. The casts were intently studied by Dr. Krantz, who was adamant that they appear to have been made by a natural animal. He reasoned that if the footprints were a hoax, then the hoaxer had to have an in-depth knowledge of anatomy.

16. Elk Wallow, Walla Walla, Washington, Paul Freeman, 1982, (3rd-generation cast, 14 inches [35.6 cm] long). The cast has an indentation in the center caused by a rock the creature stepped on. This cast is a copy of one of three casts made by Paul Freeman on which Dr. Grover Krantz discovered dermal ridges (akin to fingerprints). The chart seen below shows dermal ridge patterns of known primates compared to those found on a sasquatch footprint cast.

DERMAL RIDGES

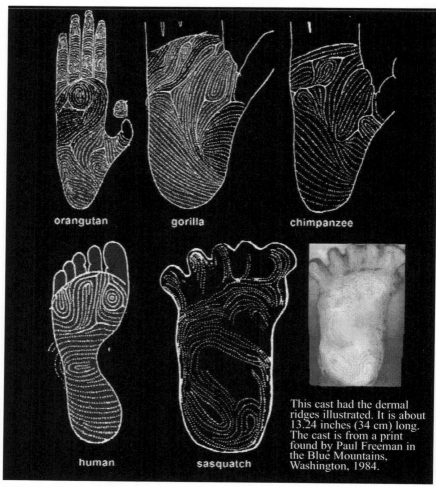

orangutan

gorilla

chimpanzee

human

sasquatch

This cast had the dermal ridges illustrated. It is about 13.24 inches (34 cm) long. The cast is from a print found by Paul Freeman in the Blue Mountains, Washington, 1984.

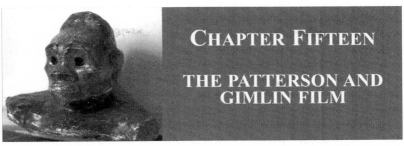

CHAPTER FIFTEEN

THE PATTERSON AND GIMLIN FILM

(This presentation was created by Christopher Murphy and Todd Prescott.)

On October 20, 1967 Roger Patterson and Robert (Bob) Gimlin of Yakima County, Washington, filmed what is believed to be a sasquatch or bigfoot at Bluff Creek, California. Their one minute (953-frame) film has become one of the most controversial films in the world. It has been debated by scientists and other professionals since October 26, 1967 and continues to remain a mystery. Absolutely nothing proves the film to be a fabrication, and nothing to date has been established beyond a doubt that the subject filmed is real.

The assertion that such a being exists (now called sasquatch or bigfoot) predates the settlement of North America by Europeans and others. Aboriginal people depicted them in their artwork and handed down stories of their

Artwork created by Michael Rugg depicting the moment Patterson and Gimlin spotted the sasquatch. (© Michael Rugg)

A scale model of the film site. The scene shows the point at which the image of the sasquatch shown on page 114 was taken. (© C. Murphy)

113

existence through generations. They have over 100 names for them, each meaning a large, hairy, ape-like being.

Reports of the entity by non-Native people emerged in the 1700s, and the total number of sightings and the finding of large footprints is now over 4,000.

Roger Patterson became intrigued with the numerous stories in the late 1950s. He went on expeditions hoping to see one of the beings or discover footprints. He wrote a book on his findings, *Do Abominable Snowmen of America Really Exist?* He then decided to make a documentary using a 16 mm movie camera.

A film frame close-up showing the subject as it turned and looked at Patterson and Gimlin. (Public domain)

Reports of large footprints found on Blue Creek Mountain, California, in late August 1967 prompted him to ask his friend Bob Gimlin to join him on an expedition to that area. The two men, both experienced horsemen, went to the area with horses.

After researching Blue Creek Mountain, the men explored the nearby Bluff Creek, California area, in which footprints had been reported some ten years earlier.

Their entire trip was uneventful up to the afternoon of October 20. Upon rounding a bend in the trail they spotted a tall, hairy, ape-like being that matched the description of sasquatch or bigfoot. The artwork (previous page) by Michael Rugg shows the scene.

The subject turned (about face) and walked away; Patterson followed it on foot taking movie footage the whole time. He ran out of film as the being disappeared into the forest.

The two men followed its path on horseback but did not see it again. They returned to the film site and studied the footprints the subject left along the creek shore. The depth of the prints indicated considerable weight. They filmed the footprints and their activities (second film roll) and proceeded to make plaster casts of two of the prints (a left and right footprint) and then left to have the films shipped for developing. They could not be sure that they actually captured the subject on film and wanted to confirm this before they left the area. They planned to stay longer and try again to film a bigfoot if the film they had taken did not show anything. However, before they could receive word, torrential rain forced them to leave the area and return to Yakima, Washington.

The plaster casts made of the subject's footprints showed an actual foot size of at least 14.5 inches, and they were very wide by human standards. The image here shows the casts with a human male footprint cast (about 11.75 inches long) for comparison.

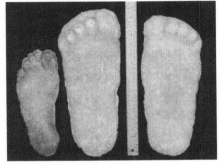

Film site casts with cast of a human foot. (© C. Murphy)

Both the film of the sasquatch and its footprints were shown to scientists at the University of British Columbia on October 26, 1967. The scientists were not allowed to express an opinion, but because further detailed study of the film was not requested it does not appear as though they were impressed.

Other scientists in the USA were consulted and their opinions varied. However, such were the result of a cursory look at the film, not a proper analysis. The first scientist to study the film in detail was Dr. John Napier in 1968. His book on the subject of sasquatch/bigfoot was published in 1972.[1]

In 1971 the film was taken to Europe for study by scientists in Finland, Sweden, Switzerland, Russia and England. Although the film was given much more attention overseas,

1. Napier, John (1972), *Bigfoot,* Berkley Publishing, New York, NY.

there were no definitive conclusions that proved the sasquatch was real, but some analysis pointed in that direction.

Two prominent Russian researchers, Dmitri Bayanov and Igor Burtsev, thoroughly studied and analyzed the film in the early 1970s and later years. They concluded that it definitely showed a living homin. Much later they published a book detailing all their findings.[2] Nevertheless, without bones or a body part, the "world of science" essentially stayed clear of the issue.

In 1975 previously unknown photographs (seen here) of three of the subject's footprints emerged. As it happened, a timber management crew was in the area three days after the filming. One of its members, Lyle Laverty, saw and photographed the prints.

Note the twig. Might this indicate a real foot made the print? (© L. Laverty)

An American 25-cent coin was placed near the big toe for size comparison. (© L. Laverty)

A smoking pipe was used for size comparison. (© L. Laverty)

These photos were taken by Lyle Laverty on October 23, 1967 of actual footprints made by the subject filmed. Six days later, October 29, 1967, casts were made of ten of the prints by Bob Titmus following up on the sighting. The Laverty photographs, however, did not emerge until 1975.

2. Bayanov, Dmitri (1997), *America's Bigfoot: Fact, Not Fiction*, Crypto-Logos Publishers, Moscow, Russia.

In that a regular 35 mm camera was used, the photos have superior clarity.

Undaunted by the lack of enthusiasm from the scientific community, Patterson proceeded to market the film and very soon "bigfoot" attained considerable notoriety. The idea that a being of this nature might inhabit the forests of North America resulted in a virtual "industry"—television productions, movies, books, and novelties. The words "sasquatch" and "bigfoot" are now household names.

It would not be until 1980 that the clearest film frames were selected and printed, with enlargements of just the subject. There was highly limited publication of this material, so only a "select few" saw all the images. There were twelve film frames printed (full frames) and twelve close-ups of just the subject. Public disclosure was not made of all images until 2004.[3] One of the close-up images is that shown on page 114.

The next scientist to evaluate the film in reasonable detail was Dr. Grover Krantz who was convinced the being was real. He also published a book on his findings in 1992.[4] Other scientists certainly reviewed the film, but did not do an in-depth analysis.

The lack of scientific involvement in the film with regard to a proper and efficient analysis bothered most researchers. In 1995 the bold step was taken to commission a forensic scientist, Jeff Glickman, to study the film and produce a detailed report on his find-

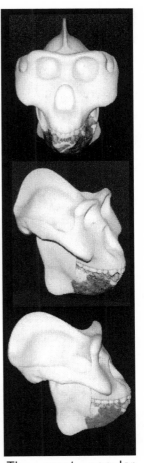

The most popular theory on sasquatch linage is that of a surviving Gigantopithecus blacki, a very large ape that inhabited Asia some 300,000 years ago. The skull shown here was constructed by Dr. Grover Krantz using an actual jaw bone for reference. (© C. Murphy)

3. Murphy, Christopher (2004), *Meet the Sasquatch*, Hancock House Publishers, Surrey BC.
4. Krantz, Grover (1992), *Big Footprints: A Scientific Inquiry into the Reality of Sasquatch*, Johnson Printing, Boulder, Co.

117

ings. His remarkable report was produced in 1998,[5] but it was not made public (printed for sale). It was eventually presented on the Internet. In summary, the report established the subject's height at 7 feet, 3.5 inches, its weight at 1,957 pounds, and its proportions beyond human standards. Many other observations resulted in the following statement by Jeff Glickman: *"Despite three years of rigorous examination by the author, the Patterson-Gimlin film cannot be demonstrated to be a forgery at this time."*

In 2014 another professional, Bill Munns, published his findings on the film.[6] He approached the issue strictly from the standpoint of determining if the "creature" was wearing a costume. Using state-of-the-art equipment, he was unable to find any indication of a costume; in fact many of his findings were to the contrary.

As to the physical film showing the subject, many copies were made of it in the late 1960s and 1970s, and it is these copies that have been used for analysis. The ORIGINAL film of the subject was put into receivership storage in Los Angeles some years after 1967. Unfortunately, the film showing footprints and other activities was copied only once or twice (as near as can be determined—only a segment showing the footprints is known) and the original was sent to England for a documentary. It does not appear it was returned and is now lost to history.

The last time the original film showing the subject was seen was in 1983 when it was taken to Hollywood, California, for analysis. Photographs had previously been made (1980) of twelve film frames. The film was borrowed from storage to do the prints and kept until 1983. It was either not returned or returned and put in the wrong storage location. Whatever the case, it too is lost to history.

The known history of the film, including the circumstances and aftermath, was detailed in a book published in

5. Glickman, Jeff (1998), *Toward a Resolution of the Bigfoot Phenomenon,* NASI.
6. Munns, William (Bill) (2014), *When Roger Met Patty,* CreateSpace Independent Publishing Platform.

2008.[7] The work was created to address the many questions regarding the film and Patterson and Gimlin. Complications (court cases) as to film ownership came to a head after Roger Patterson died (1972). His death, at age 39, was the result of Hodgkin's disease. He never wavered on his recount of the filming events and took a great amount of criticism and ridicule for this. When the subject was first spotted, he grabbed his camera rather than his rifle. However, Gimlin also had a rifle and "covered" his friend as he ran after the subject. When Patterson was terminally ill in 1972, he told a friend, "We should have shot the thing and then people would believe us." Nevertheless, the two men had a pact that they would not shoot a bigfoot unless their lives were in danger. The film reveals that the only "danger" the subject showed was a very stern look—it just calmly and intently walked away from the men.

Bob Gimlin has also suffered criticism and ridicule; so much so that for many years he did not talk about the event. In recent years he has participated in conferences and documentaries. He is about the most sincere and genuine person one could hope to meet. Now aged 85, he is still highly active and most personable.

Perhaps the most intriguing question is: Why has the film persisted? Why does it get so much attention and "heated" discussion? At this point in time, the most critical question with those who believe in the subject's reality is what kind of a being is seen in the film: non-human or human? If it is proven to be the former, then some kind of unrecognized ape inhabits North America; with the latter then we appear to have a very primitive human—perhaps providing greater insights into human evolution. In both cases, its discovery (proof of existence) would be highly significant, but more so for a human connection. To most people, news of either would simply be another news item

7. Murphy, Christopher L. (2008), *Bigfoot Film Journal,* Hancock House Publishers, Surrey BC.

(the world would not stand still). The world of science, however, would need to make some corrections ranging from, "We were wrong" (wild apes do live in North America) to "We were VERY wrong" (we are not the only "humans" on the planet). The ramifications beyond that point are left to the reader.

Although stories and artwork dealt with what we now call sasquatch or bigfoot for probably hundred of years, a color movie film that cannot be written off as a hoax is a completely different situation.

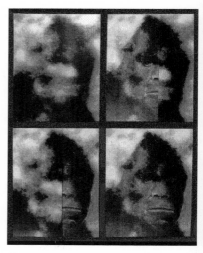

A reasonable interpretation of what the subject filmed actually looked like is this artistic rendering by Chris Murphy. It was created in 1996 using a color photocopy of the head as seen in one of the film frames (the frame previously presented). Pastels were used to reinforce what could be seen. The subject's mouth in the actual frame is partially open, so this was changed to a closed mouth to provide a more natural and aesthetically pleasing appearance. The final image has been used in many publications over the past 20 years and is likely the most publicized artwork of the subject.
(© C. Murphy)

With all of this in mind, it can be seen that Patterson and Gimlin took far more than a simple movie film of "something" on a creek shore. Their film broke through the barriers of preconceived scientific notions raising numerous questions. Such would not be the first time in history that this has happened. There are many examples of "science being wrong." However, in this case the stakes are much higher, especially (again) if the creature is human.

Fifty years is a very long time for something like the Patterson/Gimlin film to "hang in the balance," as it were. When first viewed by researchers it was thought that it would be only a matter of weeks, at the most months, before a sasquatch was found and classified. Remarkably, this was not the case and many of the early researchers have passed away.

We can thank Roger Patterson and Bob Gimlin for providing us with an enduring mystery that has become a great source of pleasure and intrigue for many, many people.

CHAPTER SIXTEEN

AUTHORITATIVE CONCLUSIONS ON THE PATTERSON AND GIMLIN FILM

Over the last 50 years, the Patterson and Gimlin film has undergone rigorous examination by highly professional and dedicated people. The following are the conclusion reached by these people all of whom studied an actual 16 mm copy of the film. *(The reports were collected and provided by Christopher L. Murphy.)*

DMITRI BAYANOV AND IGOR BURTSEV
RUSSIAN HOMINOLOGISTS

The following is a reprint from the book, *America's Bigfoot: Fact Not Fiction*, by Dmitri Bayanov, Crypto Logos Publishers, Moscow, Russia, 1997, pp. 156–158.

Conclusion

We have subjected the film to a systematic and multifaceted analysis, both in its technical and biological aspects. We have matched the evidence of the film against the other categories of evidence and have tested the subject with our three criteria of distinctiveness, consistency, and naturalness. The film has passed all our tests and scrutinies. This gives us ground to ask: Who other than God or natural selection is sufficiently conversant with anatomy and biomechanics to "design" a body which is so perfectly harmonious in terms of structure and function?*

Dmitri Bayanov

*I have deliberately phrased this sentence after one in Napier's book, *Bigfoot,* 1972.

Igor Burtsev

The Patterson–Gimlin film is an authentic documentary of a genuine female hominoid, popularly known as Sasquatch or Bigfoot, filmed in the Bluff Creek area of northern California not later than October 1967.

Until October 1967, we had lots of information on relict hominoids but they remained inaccessible to the investigators' sense of vision. We were dealing then with the underwater part of the "iceberg," as it were. October 1967 was the time when the fog cleared and the tip of the iceberg came into view. True, we still can't touch or smell this "tip," and have to be content with viewing it in the film and photographs obtained from the film. But in this we are not much different from the physician who studies a patient's bones without ever meeting the particular patient—just looking at the x-rays; or from the geologist, who studies the geology of Mars by looking at the photographs of its surface.

The difference is of course that in the geologist's case seeing is believing and, besides, he has all the might of modern science at his disposal. Those photographs cost a couple of billion dollars and nobody dares to treat them frivolously. The Sasquatch investigator, on the other hand, offered his photographic evidence to be studied by science for free and the evidence was not taken seriously.

According to Dr. Thorington of the Smithsonian, "...one should demand a clear demonstration that there is such a thing as Bigfoot before spending any time on the subject." If by a clear demonstration Dr. Thorington means a live Bigfoot be brought to his office, then it would be more of a sight for a layman than for the discriminating and analytical mind of a scientist.

Relict hominoid research is of special, potentially unlimited value for science and mankind. Thanks to the progress of the research, we know today that manlike bipedal primates, thought long extinct, are still walking the Earth in the second half of the 20th century. We also know how such a biped looks and how it walks, this knowledge being available now to any-one who wants to use their eyes.

We are indebted for this breakthrough to the late Roger Patterson, who filmed a relict hominoid in northern California in 1967, but who, to our sorrow, was not destined to witness the full triumph of his achievement.

People readily believe photographs taken on the moon, but many do not believe the Patterson and Gimlin film taken here on Earth, showing something of incalculable value for science. They do not believe it because Patterson and his assistant, Bob Gimlin, were men with no academic authority to back their claim.

And so, René Dahinden stepped forth and traveled to Moscow with his own hard-earned money to have the film analyzed and appraised in a scientific manner.

This has been done and the result is presented in this paper. The marriage of Russian theory and American practice in hominology has proven to be happy and fertile. By joining forces, we have established not only the authenticity of the film, but also that the Sasquatch is part of the natural environment of North America, and its most precious part at that. May we offer this conclusion as our modest contribution to the cause of friendship and cooperation between the peoples of the Soviet Union and North America.

The search for humanity's living roots is a cause for all mankind and this makes us look forward to new international efforts in this intriguing investigation.

The success of this research is a triumph of broad-mind-edness over narrow-mindedness and serves as an example to the world at large, which seems to be in dire need of such a lesson.

March 1977

DR. DMITRI D. DONSKOY, CHIEF OF THE CHAIR OF BIOMECHANICS AT THE USSR CENTRAL INSTITUTE OF PHYSICAL CULTURE, MOSCOW

The following is reprinted from the book, *Bigfoot/Sasquatch: The Search for North America's Incredible Creature,* by Don Hunter with René Dahinden, McClelland & Stewart Inc., Toronto, Ontario, Canada, 1993, pp. 201–204.

Qualitative Biomechanical Analysis of the Walk of the Creature in the Patterson Film

Dr. Dmitri Donskoy
(© I. Burtsev)

As a result of repeated viewings of the walk of the two-footed creature in the Patterson film and detailed examination of the successive stills from it, one is left with the impression of a fully spontaneous and highly efficient pattern of locomotion shown therein, with all the particular movements combined in an integral whole which presents a smoothly operating and coherent system.

In all the strides the movement of the upper limbs (they can be called arms) and of the lower limbs (legs) are well coordinated. A forward swing of the right arm for example, is accompanied by that of the left leg, which is called crosslimb coordination and is a must for man and natural for many patterns of locomotion in quadrupeds (in walking and trotting, for instance).

The strides are energetic and big, with the leg swung forward. When man extends the leg that far he walks very fast and thus overcomes by momentum the "braking effect" of the virtual prop which is provided by the leg put forward. Momentum is proportional to mass and speed, so the more massive the biped the less speed (and vice versa) is needed to overcome the braking effect of legs in striding.

The arms move in swinging motions, which means the muscles are exerted at the beginning of each cycle after which they relax and the movement continues by momentum. The character of arm movements indicates that the arms are massive and the muscles strong.

125

After each heel strike the creature's leg bends, taking on the full weight of the body, and smoothes over the impact of the step acting as a shock-absorber. During this phase certain muscles of the legs are extended and become tense in preparation for the subsequent toe-off.

In normal human walk such considerable knee flexion as exhibited by the film creature is not observed and is practiced only in cross-country skiing. This characteristic makes one think that the creature is very heavy and its toe-off is powerful, which contributes to rapid progression.

In the swinging of the leg, considerable flexion is observed in the joints, with different parts of the limb lagging behind each other: the foot's movement is behind the shank's which is behind the hip's. This kind of movement is peculiar to massive limbs with well relaxed muscles. In that case, the movements of the limbs look fluid and easy, with no breaks or jerks in the extreme points of each cycle. The creature uses to great advantage the effect of muscle resilience, which is hardly used by modern man in usual conditions of life.

The gait of the creature is confident, the strides are regular, no signs of loss of balance, of wavering or any redundant movements are visible. In the two strides during which the creature makes a turn to the right, in the direction of the camera, the movement is accomplished with the turn of the torso. This reveals alertness and, possibly, a somewhat limited mobility of the head. (True, in critical situations man also turns his whole torso and not just head alone.) During the turn the creature spreads the arms widely to increase stability.

In the toe-off phase the sole of the creature's foot is visible. By human standards it is large for the height of the creature. No longitudinal arch typical of the human foot is in view. The hind part of the foot formed by the heel bone protrudes considerably back. Such proportions and anatomy facilitate the work of the muscles which make standing postures possible and increase the force of propulsion in walking. Lack of an arch may be caused by the great weight of the creature.

The movements are harmonious and repeated uniformly from step to step, which is provided by synergy (combined operation of a whole group of muscles).

Since the creature is man-like and bipedal, its walk resembles in principle the gait of modern man. But all the movements indicate that its weight is much greater, its muscles especially much stronger, and the walk swifter than that of man.

Lastly, we can note such a characteristic of the creature's walk, which defies exact description, as expressiveness of movements. In man this quality is manifest in goal-oriented sporting or labour activity, which leaves the impression of the economy and accuracy of movements. This characteristic can be noted by an experienced observer even if he does not know the specifics of given activity. "What need be done is neatly done" is another way of describing expressiveness of movements, which indicates that the motor system characterized by this quality is well adapted to the task it is called upon to perform. In other words, neat perfection is typical of those movements which through regular use have become habitual and automatic.

On the whole, the most important thing is the consistency of all the above mentioned characteristics. They not only simply occur, but interact in many ways. And all these factors taken together allow us to evaluate the walk of the creature as a natural movement without any signs of artfulness which would appear in intentional imitations.

At the same time, despite all the diversity of human gaits, such a walk as demonstrated by the creature in the film is absolutely non-typical of man.

DR. D.W. GRIEVE, READER IN BIOMECHANICS, ROYAL FREE HOSPITAL SCHOOL OF MEDICINE, LONDON, ENGLAND

The following is reprinted from the book, *The Search for Big Foot, Monster, Myth or Man?* by Peter Byrne, Pocket Books, New York, N.Y., USA, 1976, pp. 137–144.

Report on the Film of a Proposed Sasquatch

The following report is based on a copy of a 16 mm film taken by Roger Patterson on October 20th, 1967, at Bluff Creek, northern California which was made available to me by René Dahinden in December 1971. In addition to Patterson's footage, the film includes a sequence showing a human being (height 6 ft., 5 1/2 in [196.9 cm] walking over the same terrain.

The main purpose in analyzing the Patterson film was to establish the extent to which the creature's gait resembled or differed from human gait. The basis for comparison were measurements of stride length, time of leg swing, speed of walking and the angular movements of the lower limb, parameters that are known for man at particular speeds of walking.[1] Published data refer to humans with light footwear or none, walking on hard level ground. In part of the film the creature is seen walking at a steady speed through a clearing of level ground, and it is data from this sequence that has been used for purposes of comparison with the human pattern. Later parts of the film show an almost full posterior view, which permits some comparisons to be made between its body breadth and that of humans.

The film has several drawbacks for purposes of quantitative analysis. The unstable hand-held camera gave rise to intermittent frame blurring. Lighting conditions and the foliage in the background make it difficult to establish accurate outlines of the trunk and limbs even in unblurred frames. The subject is walking obliquely across the field of view in that part of the film in which it is most clearly visible. The feet are not sufficiently visible to make useful statements about the ankle movements. Most importantly of all, no information is available as to framing speed used.

Body Shape and Size

Careful matching and superposition of images of the so-called Sasquatch and human film sequences yield an estimated standing height for the subject of not more than 6 ft. 5 in/1.96m. This specimen lies therefore within the human range, although at its upper limits. Accurate measurements are impossible regarding features that fall within the body outline. Examination of several frames leads to the conclusion that the height of the hip joint, the gluteal fold and the finger tips are in similar proportions to the standing height as those found in humans. The shoulder height at the acromion appears slightly greater relative to the standing height (0.87:1) than in humans (0.82:1). Both the shoulder width and the hip width appear proportionately greater in the subject creature than in man (0.34:1 instead of 0.26:1; and 0.23:1 instead of 0.19:1, respectively). If we argue that the subject has similar vertical proportions to man (ignoring the higher shoulders) and has breadths and circumferences about 25 percent greater proportionally, then the weight is likely to be 50–60 percent greater in the subject than in a man of the same height. The additional shoulder height and the unknown correction that should be allowed for the presence of hair will have opposite effects upon an estimate of weight. Earlier comments[2] that this specimen was just under 7 ft. in height and extremely heavy seem rather extravagant. The present analysis suggests that Sasquatch was 6 feet, 5 inches [1.96 m] in height, with a weight of about 280 lb [127 kg] and a foot length (mean of 4 observations) of about 13.3 inches [34 cm].

Timing of the Gait

Because the framing speed is unknown, the timing of the various phases of the gait was done in terms of the numbers of frames. Five independent estimates of the complete cycle time were made from R. toe-off, L. toe-off, R. foot passing L., L. foot passing R., and L. heel strike respectively giving: *Complete cycle time* = 22.5 frames (range 21.5–23.5). Four independent estimates of the swing phase, or single support phase for the contra-lateral limb, from toe-off to heel strike, gave: *Swing phase or single support* = 8.5 frames (same in each case).

The above therefore indicates a total period of support of 14 frames and periods of double support (both feet on the ground) of 2.75 frames. A minimum uncertainty of ± 0.5 frames may be assumed.

Stride Length

The film provides an oblique view and no clues exist that can lead to an accurate measurement of the obliquity of the direction of walk which was judged to be not less than 20° and not more than 35° to the image plane of the camera. The obliquity gives rise to an apparent grouping of left and right foot placements which could in reality have been symmetrical with respect to distance in the line of progression. The distance on the film between successive placements of the left foot was 1.20x the standing height. If an obliquity of 27° is assumed, a stride length of 1.34x the standing height is obtained. The corresponding values in modern man for 20° and 35° obliquity are 1.27 and 1.46 respectively. A complete set of tracings of the subject were made, and in every case when the limb outlines were sufficiently clear a construction of the axes of the thigh and shank were made. The angles of the segments to the vertical were measured as they appeared on the film. Because of the obliquity of the walk to the image plane of the camera (assumed to be 27°), the actual angles of the limb segments to the vertical in the sagittal plane were computed by dividing the tangent of the apparent angles by the cosine of 27°. This gave the tangent of the desired angle in each case, from which the actual thigh and shank angles were obtained. The knee angle was obtained as the difference between the thigh and shank angles. A summary of the observations is given in the table shown at right (below).

The pattern of movement, notably the 30° of knee flexion following heel strike, the hip extension during support that produces a thigh angle of 30° behind the vertical, the large total thigh excursion of 61° and the considerable (46°) knee flexion following toe-off, are features very similar to those for humans walking at high speed. Under these conditions, humans would have a stride length of 1.2x stature or more, a time of swing of about 0.35 sec., and a speed of swing of about 1.5x stature per second.

FRAME NO.	EVENT OR COMMENT	Apparent on film			Corrected for 27° obliquity		
		Thigh	Knee	Shank	Thigh	Knee	Shank
3	R. toe-off	+ 7	14	− 7	+ 8	16	− 3
4		+ 1	19	− 18	+ 1	21	− 20
5		− 7	10	− 17	− 8	11	− 19
6	blurred	− 18	3	− 21	− 20	3	− 23
7	R. foot pass L.	UNCERTAIN					
8		OF					
9		LIMB					
10		OUTLINES					
11 }	R. heel strike	HERE					
12 }		− 27	13	− 40	− 30	13	− 43
13	L. toe-off	− 25	22	− 47	− 28	22	− 50
14		0	61	− 61	0	64	− 64
15		+ 10	63	− 53	+ 11	67	− 56
16	L. foot pass R.	+ 10	64	− 54	+ 11	68	− 57
17		+ 13	62	− 49	+ 14	66	− 52
18		+ 17	45	− 28	+ 19	50	− 31
19		+ 23	38	− 15	+ 25	41	− 16
20		+ 28	29	− 1	+ 31	32	− 1
21 }	L. heel strike	+ 17	6	+ 11	+ 19	7	+ 12
22 }		+ 20	10	+ 10	+ 22	11	+ 11
23		+ 19	16	+ 3	+ 21	18	+ 3
24 }	R. toe-off	+ 17	18	− 1	+ 19	20	− 1
25 }		+ 19	33	− 14	+ 21	36	− 15
26		+ 8	15	− 7	+ 9	16	− 7
27		+ 2	19	− 17	+ 2	21	− 19
28 }	R. foot pass L.	+ 4	28	− 24	+ 4	30	− 26
29 }		NO MEASUREMENT					

Conclusions

The unknown framing speed is crucial to the interpretation of the data. It is likely that the filming was done at either 16, 18 or 24 frames per second and each possibility is considered below.

If 16 fps is assumed, the cycle time and the time of swing are in a typical human combination, but much longer in duration than one would expect for the stride and the pattern of limb movement. It is as if a human were executing a high speed pattern in slow motion.

It is very unlikely that more massive limbs would account for such a combination of variables. If the framing speed was indeed 16 fps it would be reasonable to conclude that the metabolic cost of locomotion was unnecessarily high per unit distance or that the neuromuscular system was very different to that in humans. With these considerations in mind it seems unlikely that the film was taken at 16 frames per second. Similar conclusions apply to the combination of variables if we assume 18 fps. In both cases, a human would exhibit very

131

little knee flexion following heel strike and little further knee flexion following toe-off at these times of cycle and swing. It is per-

	16 fps	18 fps	24 fps
Stride length approx.	262 cm.	262 cm.	262 cm.
Stride/Stature	1·27–1·46	1·27–1·46	1·27–1·46
Speed approx.	6·7 km./hr	7·5 km./hr	10·0 km./hr
Speed/Stature	0·9–1·04 sec.¹	1·02–1·17	1·35–1·56
Time for complete cycle	1·41 sec.	1·25 sec.	0·94 sec.
Time of swing	0·53 sec.	0·47 sec.	0·35 sec.
Total time of support	0·88 sec.	0·78 sec.	0·58 sec.
One period double support	0·17 sec.	0·15 sec.	0·11 sec.

tinent that subject has similar linear proportions to man and therefore would be unlikely to exhibit a totally different pattern of gait unless the intrinsic properties of the limb muscles or the nervous system were greatly different to that in man. If the film was taken at 24 fps, Sasquatch walked with a gait pattern very similar in most respects to a man walking at high speed. The cycle time is slightly greater than expected and the hip joint appears to be more flexible in extension than one would expect in man. If the framing speed were higher than 24 fps the similarity to man's gait is even more striking. My subjective impressions have oscillated between total acceptance of the Sasquatch on the grounds that the film would be difficult to fake, to one of irrational rejection based on an emotional response to the possibility that the Sasquatch actually exists. This seems worth stating because others have reacted similarly to the film. The possibility of a very clever fake cannot be ruled out on the evidence of the film. A man could have sufficient height and suitable proportions to mimic the longitudinal dimensions of the Sasquatch. The shoulder breadth however would be difficult to achieve without giving an unnatural appearance to the arm swing and shoulder contours. The possibility of fakery is ruled out if the speed of the film was 16 or 18 fps. In these conditions a normal human being could not duplicate the observed pattern, which would suggest that the Sasquatch must possess a very different locomotor system to that of man.

D.W. Grieve, M.SC., Ph.D.,
Reader in Biomechanics
Royal Free Hospital School of Medicine
London

References

1. Grieve D.W. and Gear R.J. (1966), "The relationships between Length of Stride, Step Frequency, Time of Swing and Speed of Walking for Children and Adults." *Ergonomics,* 5, 379–399; Grieve D.W. (1969), "The assessment of gait." *Physiotherapy,* 55, 452–460.

2. Green J. (1969), *On the Track of the Sasquatch* (Cheam Publishing Ltd.).

Conclusions Reached by the North American Science Institute (NASI)

Under the direction of J. (Jeff) Glickman, a certified forensic examiner, the North American Science Institute (NASI, now defunct) performed an intensive computer analysis on the Patterson/Gimlin film over a period of three years. At the same time, the institute carried on with general bigfoot research previously performed by The Bigfoot Research Project. In June 1998 Mr. Glickman issued a research report entitled *Toward a Resolution of the Bigfoot Phenomenon*. The report's main findings applicable to the Patterson/Gimlin film are summarized as follows:

1. Measurements of the creature:* Height: 7 feet, 3.5 inches (2.2 m); Waist: 81.3 inches (2.1 m); Chest: 83 inches (2.11 m); Weight: 1,957 pounds (886.5 kg); Length of arms: 43 inches (1.1 m); Length of legs: 40 inches (1.02m). (See Note below on height/weight.)

2. The length of the creature's arms is virtually beyond human standards, possibly occurring in one out of 52.5 million people.

3. The length (shortness) of the creature's legs is unusual by human standards, possibly occurring in one out of 1,000 people.

4. Nothing was found indicating the creature was a man in a costume (i.e., no seam or interfaces).

5. Hand movement indicates flexible hands. This condition implies that the arm would have to support flexion in the hands. An artificial arm with hand movement ability was probably beyond the technology available in 1967.

*Measurements of the arms and legs are not applicable for intermembral index calculations because they went to the finger tips and sole, not the wrist and ankle.

6. The Russian finding on the similarity between the foot casts and the creature's foot was confirmed.

7. Preliminary findings indicate that the forward motion part of the creature's walking pattern could not be duplicated by a human being.

8. Rippling of the creature's flesh or fat on its right side was observed indicating that a costume is highly improbable.

9. The creature's feet undergo flexion like a real foot. This finding eliminates the possibility of fabricated solid foot apparatus. It also implies that the leg would have to support flexion in the foot. An artificial leg with foot movement ability was probably beyond the technology available in 1967.

10. The appearance and sophistication of the creature's musculature are beyond costumes used in the entertainment industry.

11. Non-uniformity in hair texture, length, and coloration is inconsistent with sophisticated costumes used in the entertainment industry.

Mr. Glickman closes his scientific findings with the following statement:

"Despite three years of rigorous examination by the author, the Patterson–Gimlin film cannot be demonstrated to be a forgery at this time."

The full report, *Toward a Resolution of the Bigfoot Phenomenon,* is posted at the following link:

<http://www.sasquatchcanada.com/uploads/9/4/5/1/945132/rp tcol2.pdf>.

You may alternately go to the Sasquatch Canada website <http://www.sasquatchcanada.com/> and page down to the report.

CONCLUSIONS REACHED BY DR. GROVER S. KRANTZ
ANTHROPOLOGIST, WASHINGTON STATE UNIVERSITY

The following is from *Bigfoot/Sasquatch Evidence* by Dr Grover Krantz, Hancock House Publishers, 1999, pp. 122–124.

Current Status

No matter how the Patterson film is analyzed, its legitimacy has been repeatedly supported. The size and shape of the body cannot be duplicated by a man, its weight and movements correspond with each other and equally rule out a human subject; its anatomical details are just too good. The world's best animators could not match it as of the year 1969, and the supposed faker died rather than make another movie. In spite of all this, and much more, the Scientific Establishment has not accepted the film as evidence of the proposed species. There are several reasons for this reluctance that are worthy of some discussion.

Grover Krantz
(© MH Photo Library)

Most of the analyses of the film and its background were made by laymen; their studies and conclusions were published in popular magazines and books, not scientific journals. Most of these investigators did not know how to write a scientific paper or how to get one published. If they had submitted journal articles, these probably would have been rejected simply because the subject was not taken seriously by the editors, no matter how well the articles may have been written. Thus the potentially concerned scientists were simply unaware of the great quantity and quality of evidence. Most of them had heard about the movie, but were reluctant to look into it until someone else verified it. Since they all took this attitude, preferring not to risk making themselves look foolish, nothing much ever happened.

Patterson's was the first movie film ever produced purporting to show a sasquatch in the wild. Since that time many more films have appeared. I have seen eight of them and they are all fakes. A few of the most absurd of these are available on a video cassette. (One other shows a distant, non-moving object that could be a sasquatch, but there is no way to find out for sure.) Given that such faking exists now, it is not surprising that scientific interest in supposed sasquatch movies is even less today than it was back in 1967.

In many popular publications about the sasquatch there are claimed connections with the truly paranormal, and even fewer scientists want to deal with this. The lunatic fringe has the sasquatch moving through space–time warps, riding in UFOs, making telepathic connections, showing superior intelligence, and the like. All of these enthusiasts try to capitalize on anything new that comes out on the subject. Most of them will eagerly latch on to any scientist who shows an interest, and attempt to lead him/her down their own garden path. It is tantamount to academic suicide to become associated with any of these people.

Finally, and most important, there is the absence of any definitive proof that the sasquatches exist at all. If this had been a known species, the Patterson film would have been accepted without question. But without the clear proof that biologists are willing to accept, a strip of film is of little persuasive value. Of course a film like this would have been accepted as fairly good evidence for a new species of cat or skunk, but even then the type specimen would still have to be collected to make it official. For something so unexpected (at least to science) as the sasquatch, the degree of proof that is required rises proportionally.

What is said here about scientific ignorance regarding the Patterson film is equally true for the footprint evidence and the testimony of eyewitnesses. None of this is normally published in the scientific journals, hoaxes do occur, and the lunatic fringe is all over the place. I don't know of a single scientist who has firmly denied the existence of the sasquatch on the basis of a reasonable study of the evidence. Instead of this, most scientists deny it because, to the best of their knowledge, there is no substantial body of evidence that can be taken seriously.

Some of the Russian investigators, not part of their Scientific Establishment, have pushed hard for further study of the Patterson film. Their hope is that such work might establish the existence of these creatures without the necessity of collecting a specimen directly. I wish this were true. Scientific knowledge of the mechanics of bodily motion certainly has advanced in the last twenty years since Donskoy and Grieve studied the film. There are experts in sports, medicine, anatomy, athletics, running shoe design, special effects, and prosthetics who could probably make informed judgments on this film. Dmitri Bayanov has urged me and others to pursue these experts, but what efforts have been made along this line have produced no useful results. I can't afford another full round of expert-chasing after my episode with the dermal ridges, but at least I have tried.* Perhaps someone else will pursue this more diligently in the future. It is not likely that further study of the film can extract any more information than I already have, but it would make an enormous difference if a neutral expert with more appropriate credentials could just confirm what has been presented here.

(See page 112 for photographs of dermal ridges.)*

NOTE: Mr. J. Glickman, a neutral expert with appropriate credentials, did essentially confirm Dr. Krantz's findings as previously presented (NASI Conclusions). The only contentious issues were the creature's height and weight calculated by Glickman.

IMPORTANT: The conclusions reached by Dr. Esteban Sarmiento in his 2002 report (*Know the Sasquatch*, pp. 94–99) and Dr. David Daegling, *Bigfoot Exposed,* 2004 have not been included because of recent findings provided by William Munns in his book *When Roger Met Patty,* 2014.

Additional Authoritative Web Resources on the Patterson/Gimlin Film and Hominology

Footprint Analysis by Dr. Jeff Meldrum:
<http://www2.isu.edu/~meldd/fxnlmorph.html>.
Ichnotaxonomy of Giant Hominoid Tracks in North America, by Dr. Jeff Meldrum:
<http://www.sasquatchcanada.com/uploads/9/4/5/1/945132/meldrum_paper_-_ichnotaxonomy.pdf>.
The Patterson and Gimlin Film: What Makes a "Hoax" Absolutely Genuine, by Barry Keith:
<http://www2.isu.edu/rhi/pdf/Keith_rev.pdf>.
Analysis Integrity of the Patterson and Gimlin Film Images, by Bill Munns and Dr. Jeff Meldrum:
<http://www2.isu.edu/rhi/pdf/ANALYSIS%20INTEGRITY%20OF%20THE%20PATTERSON-GIMLIN%20FILM%20IMAGE_final.pdf>.
Surface Anatomy and Subcutaneous Adipose Tissue Features in the Analysis of the Patterson and Gimlin Film Hominoid, by Bill Munns and Dr. Jeff Meldrum:
<http://www2.isu.edu/rhi/pdf/Munns-%20Meldrum%20Final%20draft.pdf>.
The Patterson and Gimlin Film – Some Noteworthy Insights, by Christopher Murphy:
<http://www2.isu.edu/rhi/pdf/Murphy_PGFilmInsights.pdf>.
The Patterson and Gimlin Film Footage:
<https://www.youtube.com/watch?v=v77ijOO8oAk>.
<http://www.relativelyinteresting.com/heres-what-the-bigfoot-patterson-gimlin-film-looks-like-when-its-stabilized/>.
Historical Evidence for the Existence of Relict Hominoids, by Dmitri Bayanov:
<http://www2.isu.edu/rhi/pdf/Bayanov_rev.pdf>.
The Almasty of the Caucasus – Life Style of a Hominoid, by Dr. Marie-Jeanne Koffmann:
<http://www2.isu.edu/rhi/pdf/Koffmann_2.pdf>.
The Sierra Sounds:
<http://www.sasquatchcanada.com/uploads/9/4/5/1/945132/sierra_sounds_rev_aug6-17_pdf.pdf>.
Website Links:
Relict Hominoid Inquiry <http://www2.isu.edu/rhi/>.
Sasquatch Canada <http://www.sasquatchcanada.com/>.

CHAPTER SEVENTEEN

A CALL TO ACTION

This entire Chapter was provided by Christopher L. Murphy. The intention is to leave the reader with a synopsis of the current situation and point out the importance of the material presented in this book.

To achieve the objective of having hominology recognized as a valid scientific discipline the first requirement is to bring about cooperation in our badly dysfunctional "fraternity" and restore credibility to the homins we believe exist (sasquatch/bigfoot, Russian snowman/almasty, yeti, yowie and yeren). All have become "objects of entertainment" for the purpose of making money. Even skepticism is used to "make a buck."

Unfortunately, the "world of science" wants no part of tomfoolery and individual scientists are cautious to stay clear of the subject even if they harbor an interest. Checking the word "bigfoot" on Google will show 24,600,000 results, so there is no shortage of interest, including use of the name; it's just not the RIGHT interest. There is no intention here to belittle the few scientists we have; their work is invaluable.

The establishment of the Relict Hominoid Inquiry under Idaho State University was, and continues to be, the most significant step towards homin recognition. It is the only website with scientific credentials. Other websites are essentially like newspapers or magazines. They provide information that is entertaining, but little is looked at by professionals (although I have no way of knowing that for certain).

What we have here is a bit of a Catch-22 situation. If scientists don't know what is involved in hominology, then how can we expect them to support research? They shy away from getting knowledge for fear of what has become known as the "curse of bigfoot."

So, what would change if hominology somehow got the "green light?" In other words it would be okay for scientists to explore the subject. I don't think we would be inundated with request for information, but there would definitely be an increase in professional attention to the subject. There would be articles in scientific journals and universities would consider publishing books that present the possibility of homin existence.

With the door now open, as it were, and professional people getting involved in the subject the desire to resolve the issue with tangible conclusive evidence would greatly increase. This would lead to funding for research—even some expeditions to remote areas to have a look at what people have been saying for centuries.

Most certainly, at this juncture the first thing to do is get this work out to major scientific organizations and universities and thereby put the case on the table. Also, to get as much exposure as possible through the usual media outlets (including websites). Readers are asked to assist here in any way they can. Please recommend this book to others, discuss it on forums, and reference it in articles and papers.

Keep in mind that the objective is scientific recognition of hominology to eventually bring about a resolution to the age-old question of homin existence. In particular, the thousands of witnesses who "know what they saw" will get great satisfaction when the question is answered in the affirmative (as we believe it will).

Of great help would be the formation of a formal international society for research in hominology. This could be much the same as the defunct International Society of Cryptozoology (ISC). The difference being that only hominoids would be studied. Preferably the new society would be hosted by a university or major research institution like the Smithsonian or

National Geographic. Its board of directors would need to be accredited scientists. Funding would need to be handled by tax-deductible dues. A website would be necessary for the presentation of formal papers, and perhaps a printed annual journal could be provided.

At the outset of this book your attention was drawn to the many decades of ardent research performed in North America on the sasquatch/bigfoot question—obviously we need help; what we have done and are doing is simply not enough to get proper scientific involvement in hominology. *The Making of Hominology* is a major step in moving toward a resolution.

BIBLIOGRAPHY

Bauer, Henry, (2001). *Science or Pseudoscience.* Urbana and
    Chicago, USA: University of Illinois Press.
———, (2017). *Science Is Not What You Think.* Jefferson,
    North Carolina, USA: McFarland & Company.
Bayanov, Dmitri, (1991) *Wood Goblin Dubbed "Ape": A
    Comparative Study In Demonology* (in Russian).
    Moscow, Russia: Society to Study Mysteries and
    Enigmas of the Earth.
———, (1996). *In the Footsteps of the Russian
    Snowman.* Moscow, Russia: Crypto-Logos
    Publishers.
———, (1997). *America's Bigfoot: Fact, Not Fiction: US
    Evidence Verified in Russia.* Moscow, Russia:
    Crypto-Logos Publishers.
———, (2001). *Bigfoot: To Kill or to Film: The Problem of
    Proof.* Burnaby, BC, Canada: Pyramid Publications.
———, (2009 to 2011). *"Learning from Folklore."* A series
    of papers originally published on the Bigfoot
    Encounters website; currently on the Sasquatch
    Canada website.
———, (2011). *Bigfoot Research: The Russian Vision.*
    Surrey, BC, Canada: Hancock House Publishers.
———, (2018). "Is A Greater Paradigm Shift Thinkable?"
    Paper published on the Sasquatch Canada website.
Bernheimer, Richard (1952). *Wild Men in the Middle Ages.*
    Cambridge, USA: Harvard University Press.
Byrne, Peter (1976). *The Search for Big Foot: Monster, Myth
    or Man?* New York, NY, USA: Pocket Books.

Glickman, Jeff (1998). *Toward a Resolution of the Bigfoot
    Phenomenon.* Mt. Hood, Oregon, USA: North
    American Science Institute.
Green John (1969), *On the Track of the Sasquatch.* Agassiz,
    BC, Canada: Cheam Publishing Ltd.

————, (1978). *Sasquatch: The Apes Among Us.* Surrey, BC, Canada: Hancock House Publishers.

Healy, Tony, and Cropper, Paul (2006). *The Yowie: In Search of Australia's Bigfoot.* Sydney, Australia: Strange Nation.

Hunter, Don and Dahinden, René (1993) *Bigfoot/Sasquatch: The Search for North America's Incredible Creature.* Toronto, Canada: McClelland & Stewart Inc.

Krantz, Grover (1992), *Big Footprints: A Scientific Inquiry into the Reality of Sasquatch.* Boulder, Colorado, USA: Johnson Printing.

————, (1999). *Bigfoot Sasquatch Evidence.* Surrey, BC, Canada: Hancock House Publishers.

Kuhn, Thomas (1962). *The Structure of Scientific Revolutions.* Chicago, USA: University of Chicago Press.

Markotic, Vladimir (editor) and Grover Krantz (associated editor) (1984). *The Sasquatch and other Unknown Hominoids.* Calgary, Alberta, Canada: Western Publishers.

Meldrum, D.J. (1999). Evaluation of Alleged Sasquatch Footprints and Inferred Functional Morphology. *American Journal of Physical Anthropology;* Supplement 27:161 (Abstract).

————, (2006). *Sasquatch: Legend Meets Science.* New York, NY, USA: Tom Doherty Associates.

Morehead, Ronald J. (2013). *Voices in the Wilderness* (Second Edition). California, USA: Self-Published.

Moskowitz Strain, Kathy (2008). *Giants, Cannibals & Monsters: Bigfoot in Native Culture.* Surrey, BC, Canada: Hancock House Publishers.

Munns, William (2014). *When Roger Met Patty.* USA: CreateSpace, Independent Publishing Platform.

Murphy, Christopher (2004). *Meet the Sasquatch.* Surrey, BC, Canada: Hancock House Publishers.

————, (2010). *Know the Sasquatch/Bigfoot.* Surrey, BC, Canada: Hancock House Publishers.

————, (2008). *Bigfoot Film Journal.* Surrey BC, Canada: Hancock House Publishers.

Napier, John (1973). *Bigfoot.* New York, USA: Dutton Publishers.

Roosevelt, Theodore (1893). *Wilderness Hunter: Outdoor Pastimes of an American Hunter.* New York, NY, USA: G. P.   Putnam & Sons.

Sanderson, Ivan (1961). *Abominable Snowmen: Legend Come to Life.* Kempton, Illinois, USA: Adventures Unlimited Press (2006 edition).

Shackley, Myra (1983). *Still Living? Yeti, Sasquatch and the Neanderthal Enigma.* New York, NY, USA: Thames and Hudson.

Sprague, Roderick and Krantz, Grover – editors (1977). *The Scientist Looks at the Sasquatch.* Moscow, Idaho, USA: University Press of Idaho.

Wilson, Thomas (2015). *Bigfoot in Evolutionary Perspective: The Hidden Life of a North American Hominin.* USA: CreateSpace, Independent Publishing Platform.

GENERAL INDEX

Abbott, Don, 8, 103

Abbott Hill (Washington), 108, 109

Agnagna, Marcellin, 74

American Museum of Natural History, 49

Bauer, Henry, 18, 31, 45, 46, 68

Bauman (trapper), 15

Bayanov, (Front), 2, 3, 4, 5, 6, 7, 10, 11, 12, 13, 19, 55, 61, 62, 66,67, 68, 116, 122, 124, 138, 139

Beebe, Frank, 8, 9

Belshaw, Cyril, 65

Bernheimer, Richard, 43

BFRO, 81

Bigfoot Field Researchers Organization, 81

Bigfoot Research Project, 134

Bigfoot Times, 81

Bindernagel, 6, 18, 19, 36, 83, 86, 89, 109,

Biot, Jean-Baptise, 33, 34

Blue Creek Mountain, California, 103, 108, 114

Bluff Creek, California, 67, 102, 108, 110, 111, 113, 114, 123, 128

Bonch-Osmolovsky, G.A., 58

Bossburg, Washington, 104, 111

Bounak, V.V., 58

Bourtsev, Igor (see Burtsev)

Bourtseva, Alexandra, 5, 6

Brehm, Alfred, 14

British Columbia Museum (now Royal Museum), 103

Brown University, 84

Bruno, (Giordano), 10, 34

Buckley, Archie, 90

Burke, John G., 32

Burtsev, Igor, 6, 52, 55, 61, 62, 65, 66, 68 75, 95, 116, 122, 124, 125

Butler, Jim, 91

Byrne, Peter, 6, 61, 62, 128

Carl, Clifford, 8

Carter, Janice, 85, 86

Central Scientific Research Institute of Prosthetics and Artificial Limb Construction, 52

Chladni, Ernst Florens, 33, 34, 35

Chomsky, Noam, 17

Clinton, Bill (President), 98, 99

Coleman, Loren, 83

Columbia University Press, 97

Committee on Cinematography (Moscow, Russia), 53, 56

Coon, Carlton, 67

Crew, Jerry, 108, 110

Crowe, Ray, 92

Cryptozoology (journal), 70, 72, 82, 88

Czarina Catharina, 32

Daegling, David, 97, 138

Dahinden, René 6, 7, 9, 50, 52, 53, 55, 56, 60, 61, 62, 63, 66, 68, 87, 111, 124, 125, 128

Darnton, John, 31, 32, 40

Dart, Raymond, 51, 76

Darwin Museum, Moscow, 7, 20, 22, 58, 65, 67, 84

Darwin, Charles, 18, 42, 89

Department of Vertebrate Zoology, National Museum of Natural History, Smithsonian Institution, 71

Dermal Ridges, 76, 112, 138

146

Dinsdale, Tim, 82, 94
Do Abominable Snowmen of America Really Exist?, 9, 114
Donskoy, Dmitri, 6, 62, 63, 87, 125, 138
Druyan, Ann, 27
Dubois, Eugene, 51, 76
Duff, Wilson, 8
Duncan, Will, 6, 92
Dyrrachium (modern Durres) Albania, 42
Edwards, York, 8
Elk Wallow, Walla Walla, Washington, 112
First Russian Report on the 1967 Bigfoot Film, 61
Forth, Gregory, 94
Freeman, Paul, 112
French Academy of Sciences, 32, 33, 48
Galilei, (Galileo), 10, 34, 42
Gear, R. J., 133
Gimlin, Robert (Bob) 6, 85, 98, 99, 102, 113, 114, 119, 120, 121
Glickman, Jeff, 117, 118, 134, 135, 138
Goodall, Jane, 9, 18
Graves, Pat, 110
Grays Harbor County, Washington, 103
Green, John, 6. 9, 20, 37, 49, 58, 79, 80, 81, 82, 83, 85, 86, 108, 113
Green, Mary, 85
Greenwell, Richard, 70, 71, 72, 73, 83
Greiner, Thomas M., 25
Grieve, D.W., 128, 132, 133, 138
Groves, Colin, 70

Guiquez, Charles, 8
Haas, George, 6. 90
Halpin, Marjorie M., 65, 66, 67
Hancock, David, 4, 8, 10, 93
Hancock House Publishers, 8, 9, 18, 81, 117, 119, 136
Hart, Haskell V., 19, 68
Heinselman, Craig, 92
Heryford, Dennis, 103, 109
Heuvelmans, Bernard, 24, 71, 76, 82, 96
Hitler, (Adolph), 14
Horgan, John, 28
Hollywood, California, 118
Hominoid Problem Seminar, Darwin Museum, 58
Howard, Edward, 33, 34
Hunter Don, 61, 62, 125
Huxley, Thomas, 45
Hyampom, California, 110
Idaho State University, 13, 16, 94, 140
Institute and Museum of Anthropology, Moscow, Russia, 52
Institute of Ethnography, Moscow, Russia, 52
International Center of Hominology, Moscow, 20
International Society of Cryptozoology, 7, 71, 80, 82, 96. 141
Johanson, Donald, 76
Joint Air Reconnaissance Intelligence Centre (Royal Air Force, Great Britain, 82
Journal of Scientific Exploration, 37, 68
Journal Znanie-Sila, 82
Keith, Barry, 139
Ketchum, Melba, 68
Koffmann, Marie-Jeanne,

(Front), 6, 7, 52, 67, 78, 139

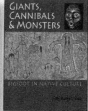

Made in the USA
Middletown, DE
11 October 2020